LOST FROGS AND HOT SNAKES

LOST FROGS AND HOT SNAKES

Herpetologists' Tales from the Field

Edited by Martha L. Crump

COMSTOCK PUBLISHING ASSOCIATES

AN IMPRINT OF CORNELL UNIVERSITY PRESS   ITHACA AND LONDON

A portion of the proceeds from this book will be donated to the Society for the Study of Amphibians and Reptiles (SSAR) Founders' Fellows Program, which supports budding high school herpetologists to attend the annual SSAR meetings. At the meetings, students attend paper sessions, learn about colleges and universities with herpetology programs, discover career opportunities, and meet others with shared interests.

First published 2024 by Cornell University Press
Printed in the United States of America

Library of Congress Cataloging-in-Publication Data

Names: Crump, Martha L., editor.
Title: Lost frogs and hot snakes : herpetologists' tales from the field / edited by Martha L. Crump.
Description: Ithaca [New York] : Comstock Publishing Associates, an imprint of Cornell University Press, 2024. | Includes bibliographical references and index.
Identifiers: LCCN 2023039433 (print) | LCCN 2023039434 (ebook) | ISBN 9781501774485 (paperback) | ISBN 9781501774492 (pdf) | ISBN 9781501774508 (epub)
Subjects: LCSH: Herpetology—Fieldwork—Anecdotes. | Herpetologists—Anecdotes. | LCGFT: Anecdotes.
Classification: LCC QL644 .L67 2024 (print) | LCC QL644 (ebook) | DDC 597.9—dc23/eng/20231101
LC record available at https://lccn.loc.gov/2023039433
LC ebook record available at https://lccn.loc.gov/2023039434

Contents

Preface

> **Perhaps, when looking back to the years and years of fieldwork, we will realize that, albeit sometimes we did not find everything we expected or cherished, for sure we always met the very best version of ourselves. Fieldwork forever!**
>
> —Ignacio De la Riva, 2021

Fieldwork is the beginning and the end for many of us. We are addicted to the thrill of discovery, the novelty of exotic landscapes and iconic species, and the allure and magic of the unknown. Fieldwork opens up whole new worlds, often exposing us to new cultures, expanding our perspectives on social and political issues, and increasing our sensitivity to diverse lifestyles, customs, and attitudes. Seeing the world through a different lens helps us to understand ourselves—who we are and who we want to be. Never mind the extreme temperatures, biting and stinging insects, cold showers (or lack thereof), bacterial and fungal ailments, and monotonous diet. We are happiest when we are in the field. It is who we are, a sentiment echoed by the contributors in this collection of essays who portray fieldwork like it really is—the discomforts, frustrations, dangers, failures, successes, inspirations, surprises, exhilarations, adventures, and discoveries.

No doubt many of us will continue doing fieldwork even when we stagger under the weight of heavy backpacks, trip over tree roots at night, and are unable to hear the high-pitched peeps of treefrogs. Perhaps our children or students will carry our heavy backpacks, lead us by the hand down treacherous trails, and encourage us to sign up for those darn hearing aids. Fieldwork is in our blood.

It is most fitting that this book be published by Comstock Publishing at Cornell University Press. Field biologists study nature in nature, and that is exactly what Anna Botsford Comstock (1854–1930) advocated as one of the first educators to encourage teachers to take their students *outside* to study nature. During the late 1800s, Comstock was a major proponent of the Nature Study

Movement, whose mantra was "study nature, not books." She found that most New York public schools did not teach nature study and that many teachers felt inadequate to teach the subject. Deciding something needed to be done, in 1909 she began writing a guide for teachers, entitled *Handbook of Nature Study*. She published the book in 1911, through her husband's publishing house, Comstock Publishing Company. The book was 938 pages, with 232 planned lessons and suggested field trips and experiments. By 1923 when the League of Women Voters voted Anna Comstock one of the twelve greatest women in America within their chosen fields, the book had already sold over forty thousand copies. The *Handbook* has gone through twenty-five editions and has been translated into eight languages. Thanks to Cornell University Press, it is still in print and is widely used by teachers, biologists, and laypersons. It is an honor to have this collection of essays in such august company, and I can only hope that the book, like Comstock's *Handbook*, serves to inspire the next generation of field biologists.

References

Comstock, A.B. 1911. Handbook of nature study. Ithaca, NY: Comstock Publishing Company.
De la Riva, I. 2021. Fieldwork forever. *Herpetological Review* 52(3): 559–563.

Map showing the geographic locations of field sites discussed in the essays. The numbers correspond with the essay numbers. This map was created by Alison Davis Rabosky.

LOST FROGS AND HOT SNAKES

INTRODUCTION

Field biologists are a unique breed of scientists who study plants and animals in their natural environments. We often appear disheveled in our baggy, quick-drying clothes and dorky fishing vests bulging with specimen vials, counters, flagging tape, tape measurers, field guides, and waterproof notebooks. Binoculars likely dangle from our sweaty necks. Our backpacks are filled with, depending on our particular field of interest, a portable plant press, newspapers, fruit and seed bags; insect sweep net with detachable handle, Berlese funnel, aspirators, specimen vials, and tweezers; mammal traps, peanut butter, oats, and leather gloves; snake tongs, plastic and cloth bags, calipers, head lamp, and extra batteries; bird bands, pliers, mist nets, and Pesola scales. Often inexplicable to our friends and family members, we tolerate mud, rain, biting and stinging insects, extreme heat and cold, and primitive living conditions.

Those of us who are field herpetologists—field biologists who focus on amphibians and reptiles—have never outgrown our childhood passions of rolling logs in hopes of finding a salamander, watching tadpoles "magically" transform into frogs, striving to outrun lizards, sneaking up on basking turtles, and attempting to catch crocodilians and snakes without getting bitten. For us, there is nothing more satisfying than spending a day with our favorite animals. We are living our dreams.

That's who we are, but what exactly is it that we do "in the field"? We study basic biology (e.g., reproduction, foraging strategies, defense mechanisms, and physiological adaptations); spend long hours watching animals behave (and often even longer hours watching them do nothing); carry out long-term monitoring

1

studies of endangered species; gather natural history data fundamental in framing ecological and evolutionary questions; perform quantitative experiments to answer ecological questions; survey previously unexplored regions in search of undescribed species; carry out biodiversity surveys for conservation and land preservation measures; and much more.

Why this book? I invited a diverse and distinguished group of field herpetologists each to share a meaningful field experience for several reasons. The essays in this collection entertain and delight, but they also offer an understanding for what field biology is, what field biologists do, and how we go about doing it. The book focuses on field herpetologists, but our experiences, and thus our stories, reflect biological fieldwork as a whole. I hope that through these stories, nonbiologists and aspiring biologists will better appreciate the value of fieldwork and the value of amphibians and reptiles we know to be so critical to the health of our planet.

I also hope these essays will inspire young readers and early career biologists to explore the possibilities of field biology. Recently, I interviewed over sixty-five women for *Women in Field Biology: A Journey into Nature* (Crump and Lannoo 2022). I was struck by how many women told me that they discovered only late in their undergraduate college years that field biology was a viable career option. Some had discovered field biology by accident through reading scientific papers and realizing, "Wow, people can earn a living by studying mate selection in dung beetles!" What better sales pitch could there be for spreading the word about field biology than the voices of those of us passionate about our careers?

And finally, I hope the book will inspire readers to become more engaged observers of the natural world. In the past several decades, numerous polls and studies carried out by natural and social scientists have documented that much of the human population feels psychologically disconnected from their natural surroundings. Many have lost their childlike sense of wonder. As individuals, families, and communities, we need to strengthen our bonds with nature, for our own mental and physical health and for the health of the environment. I hope the essays in this collection will rekindle readers' childhood memories of the joy and wonder they experienced while exploring the out-of-doors. And if so, perhaps these stories will inspire readers to engage their curiosity, go outside, and poke around in nature once again.

So, what are you about to read in this collection of essays? Allow me to offer a brief roadmap and a few teasers.

Part 1. Many of us are passionate about field biology because of the **thrill of discovery**. Robert Espinoza expresses this sentiment eloquently in the opening essay. He writes: "There's no greater high for a scientist than discovery. It's a cocktail of adrenalin and joy that imbues a pure sense of euphoria. . . . And like many drugs, it's often in short supply and addictive. Consequently, the brains of scientists crave it, so we are always on the lookout for opportunities to achieve our next dopamine-induced fix. For field biologists, there's no place more alluring than nature. For it's in the field where we feel simultaneously at ease yet invigorated, quieted yet energized. Those sensations often come from a deep appreciation—some call it love—of the organisms we study." You'll join Espinoza as he assembles puzzle pieces to reveal nine cryptic species of lizards in northern Argentina, and you'll share in the excitement as Jodi Rowley discovers "the frog that sings like a bird" on a mountain in Vietnam.

Part 2. The promise of **adventure and exploration** lures many of us to field biology. David Bickford contends with leeches, gets bumped in the dark by a Papuan forest wallaby, and discovers a new mode of parental care in male microhylid frogs while exploring Papua New Guinea. Maureen Donnelly shares her "firsts" while exploring a tepui as part of the Tapirapecó Expedition in Venezuela. Karen Lips and two field companions set off to explore Valle de Silencio, a high-elevation bog near the continental divide in Costa Rica, lured by the promise of caecilians and tapirs. The trio never make it to their destination, but for years they have laughed about "mystical places where caecilians are dripping from trees and tapirs are prancing in the forest."

Part 3. We have a **fascination and love for the animals** we study. Alan Savitzky writes about the privilege of finding a Fea's Viper on Gongshan Mountain of western China—a rarely encountered "mythical serpent." Susan Walls has been infatuated with mole salamanders since she was bamboozled by the sight of her first one 40 years ago: "an odd creature with smooth skin, four legs, and big floofy, feathery, bright red external gills." Erin Muths has studied a population of chorus frogs in the Never Summer Range, north-central Colorado, for 25+ years. She writes: "I will immerse myself in their ritualized world again this spring, and I know that they have changed me in subtle ways I may not even be aware of." Wolfgang Wüster contemplates the resiliency and fragility of life in the context of an encounter with a Western Diamondback Rattlesnake in Arizona.

Part 4. Along with the joys of fieldwork come **mishaps and misadventures**. When a protective mother alligator threatens Whit Gibbons' 12-year-old son Michael, who is holding one of her babies, one of Whit's colleagues yells, "Climb a tree!" Another yells, "Throw the baby in the lake!" Whit yells, "Run!" Kelly Zamudio steps on a Black-Tailed Rattlesnake while studying lizards in the Peloncillo Mountains, New Mexico. It is not a dry bite. Robert Hansen runs into presumed *narcotráficos* in Sinaloa, Mexico, while hunting snakes with a herping buddy. William Lamar and his herping buddies find themselves delicately balanced on a narrow road in Oaxaca, Mexico, when the right rear tire of their truck dangles over a precipice into black emptiness. In Chiapas, Mexico, an angry mob assumes that Oscar Flores-Villela and his companions are commercial collectors intending to steal a boatload of Central American River Turtles.

Part 5. Field biologists know the necessity of **dealing with the unexpected**: the twists and turns, which sometimes turn into knots and can make or break a field season. Nature is unpredictable, out of our control, and so we must be flexible. We learn to expect the unexpected. Alessandro Catenazzi's dissertation research takes a "sharply descriptive turn" after the excruciating failure of his beach wrack experiment in Paracas National Reserve, Peru. Karen Warkentin, planning to study her "escape-hatching" hypothesis of Red-Eyed Treefrog tadpoles at Corcovado Park, Costa Rica, waits for an entire month for the rains to come (and considers studying crabs instead). Alison Davis Rabosky stares at her positive pregnancy test result at the beginning of nine weeks of remote fieldwork in Western Australia and later contemplates, "Is there some kind of safe limit to how many kilometers you can drive on horrible washboard roads before you scramble your baby's developing brain?"

Part 6. In the end, sometimes it is **the people we meet, the friendships we forge, and the students we influence** who leave the strongest impressions on us. A strong, long-term bond forms between Sinlan Poo and a Taiwanese cargo ship captain incarcerated in Panama for manslaughter. Lee Fitzgerald, while working as a Peace Corps volunteer in Paraguay, films an episode of *Mutual of Omaha's Wild Kingdom* television show entitled "The Unexplored Gran Chaco" with the show's star, Marlin Perkins. A student holds a tiny, red Amazon Egg-Eater Snake and loses her lifetime fear of snakes during a field course led by Tiffany Doan in the Peruvian rainforest.

The fifty contributors are a diverse group of herpetologists from all over the world: white and people of color; museum, zoo, independent, state and federal, and academic professionals; and early career through retired individuals. Their stories range in time from the 1960s through the present, and they take place in Argentina, Australia, Brazil, Canada, China, Colombia, Costa Rica, Ecuador, Guatemala, Guinea, Japan, Madagascar, Mexico, New Caledonia, Panama, Papua New Guinea, Paraguay, Peru, Sarawak, Slovenia, Taiwan, Venezuela, Vietnam, and from the west to east coasts in the United States. The stories feature diverse habitats and landscapes: beaches, deserts, rainforests, cloud forests, temperate forests, floodplains, páramo, puna, and chaco, and the focal animals include alligators, goannas, geckos, salamanders, frogs, turtles and tortoises, vipers and rattlesnakes, and more. Our tales from the field are as much about us as the animals we study and the experiences themselves.

Collectively, these stories paint a diverse and honest collage composition of fieldwork. They reveal the value of studying animals in nature. And they showcase the charm—even seductiveness—of amphibians and reptiles, animals not always appreciated by the general public. The next time you see a frog, salamander, lizard, snake, turtle, or crocodilian in the field, stop to have a close look. You just might understand why herpetologists find these animals so fascinating and worthy of our love. And you just might deepen your own appreciation for amphibians and reptiles.

Reference

Crump, M.L., and M.J. Lannoo. 2022. Women in field biology: A journey into nature. Boca Raton, FL: CRC Press, Taylor and Francis Group.

Part I
THE THRILL OF DISCOVERY

THE IRREPLACEABLE ROLE OF NATURE IN SCIENTIFIC DISCOVERY

Robert E. Espinoza

There is no greater high for a scientist than discovery. It's a cocktail of adrenalin and joy that imbues a pure sense of euphoria. Yet because it's natural, there's no hangover on the downside. In fact, it can last and last. Even recollections of long-past discoveries can evoke these sensations. And like many drugs, it's often in short supply and addictive. Consequently, the brains of scientists crave it, so we are always on the lookout for opportunities to achieve our next dopamine-induced fix.

For field biologists, there's no place more alluring than nature. For it's in the field where we feel simultaneously at ease yet invigorated, quieted yet energized. Those sensations often come from a deep appreciation—some call it love—of the organisms we study. For many herpetologists, that passion is traceable to an early childhood experience. I grew up in Southern California and mine came in kindergarten. One day, two older boys came to our class for "show and tell" and passed around a jar containing several Pacific Chorus Frogs (*Pseudacris regilla*). I was rapt, and especially enamored with a bright green male. Yet, despite staying late to beg to take them home, my teacher patiently explained she had no authority to give them to me. So a few days later, I convinced my father to take me to a drainage canal near our home to search for my own. That evening, we captured about a dozen adult Western Toads (*Anaxyrus boreas*). I fondly remember their unique body patterns (so distinct, I named each one), the endearing bleating of the males when grasped under their armpits, and their mossy–earthen aroma. My fate was sealed. Although I had been interested in science and nature since

I was a toddler, these formative experiences with amphibians proved decisive: I would become a herpetologist.

I spent most of my youth immersed in herpetological pursuits. By the time I was an early teen, I had countless pet frogs, salamanders, turtles, lizards, and snakes (both local and exotic) and raised my own mice and rats to feed them. (Major shout out to my mom for enduring the many escapees and rodent-cicles in our freezer!) In junior high, I started attending monthly meetings of the San Diego Herpetological Society (SDHS). There, I learned about the diverse pursuits of herpetologists: anatomy, behavior, ecology, biogeography, systematics, and so on (this was prior to the takeover of most amateur herp societies by those primarily interested in breeding herps for profit). I began collecting herp books and taking notes and photographs on field excursions. I also met professional herpetologists and others who shared and encouraged my interests, including several boys my age who also aspired to be herpetologists. For several years, I hiked nearly every weekend along the river valleys, mountains, and deserts of Southern California with my SDHS pals in search of herps. Those experiences strengthened my love of nature, reinforced the knowledge acquired from field guides, and sharpened my field-herping skills. In high school, I shared my live animals and growing knowledge of herps with various groups and schools and worked in local pet stores as the herp specialist. I supported myself though college, in part, by selling lizards and snakes I bred in captivity. Immediately upon transferring from community college to San Diego State University as a college sophomore (and blissfully unaware of prerequisites), I enrolled in Dr. Richard Etheridge's senior-level herpetology class. That was the single-most important step I made toward a career in herpetology. Not so much for what I learned (although there was plenty of knowledge acquired in that rigorous course) but even more for the lifelong network of colleagues and friends established through Richard.

Indeed, it was Richard, who about seven years later, invited me to accompany him on a trip to Argentina and introduced me to the wonders of its herpetofauna and the warm and gifted herpetologists at the Fundación Miguel Lillo (FML) at the Universidad Nacional de Tucumán. It was the austral summer of 1994–1995. I had just started the doctoral program at Colorado State University and had plans to study the evolution of herbivory in reptiles. My investigation would focus on a then poorly studied group of lizards in the genera *Liolaemus* and *Phymaturus* from the southern cone of South America. These are small to moderately sized lizards (lacking fitting common names) that occupy a diversity of ecological niches from Andean Peru to nearly the southern tip of the continent and from sea level to above 5000 m. *Liolaemus* was a fairly large group, with about 150 species recognized at the time (nearly double that currently); likewise, *Phymaturus* had only 10 species (now 50).

I learned more than about herps on that trip. So much more. It was my first full immersion in a culture outside the United States. (Not counting high school jaunts for surfing, tacos, and beer in northern Baja California.) Fortunately, we had knowledgeable and patient hosts at the Instituto de Herpetología at the FML, including (then) graduate students Virginia Abdala, Félix Cruz, Monique Halloy, Fernando Lobo, Adriana Manzano, and Silvia Moro; senior herpetologists Esteban Lavilla, Ricardo Montero, and Gustavo Scrocchi; and the world-renowned herpetologist, Raymond Laurent, who served as the institute director. In January 1995, Richard, Esteban, Fernando, local undergraduate Juan Carlos Moreta, and I set out in a well-aged field truck, primarily in search of *Liolaemus*. Our trip included herping dozens of localities in seven northwestern provinces, primarily along the Andean cordillera. I was surprised by how similar the climate and many of the habitats were to those in the southwest United States. Even numerous plant genera were shared between the two regions. I found the many instances of convergence between species of Argentinian herps and those from Southern California to be remarkable. Although memorable on many levels—particularly the new herps we encountered (not to mention the 13 flat tires we endured)—the grander herpetological discoveries would come a few years later.

Back in Tucumán, in the course of dissecting lizard guts to determine the diets of *Liolaemus* in the FML collection, I often encountered females with either shelled eggs or developing embryos. After recording the reproductive modes of a few dozen species, followed by a cursory literature review, my interests detoured from herbivory to an investigation of the evolution of viviparity within the group. When I shared my new plans with Fernando, he told me about Martha Patricia Ramírez-Pinilla, a recently graduated doctoral student who had studied the reproductive cycles of several species of *Liolaemus*, including some in the *L. alticolor* group. She found that this group included both oviparous and viviparous species, but remarkably, one species (*L. alticolor* at the time) was reproductively bimodal—a species within which some females produce eggs and others give birth to live young. Martha had even identified a reproductively bimodal population of this species. This was intriguing because reproductive bimodality is very rare in reptiles, and at the time, there weren't any known reproductively bimodal populations of squamate reptiles (lizards or snakes).

So, in the early austral summer (November) of 1996, Fernando and I set out to study this bimodal population, which occurred just a couple hour's drive from the FML. Prior to my arrival, Fernando had reexamined Martha's reproductively bimodal specimens with the keen eye of a taxonomist and had identified morphological differences between the oviparous and viviparous morphs. Notably, the viviparous morphs had two dorsal fields of black-and-white paravertebral markings bordered by narrow golden-yellow stripes and speckled bellies (labeled A in the following figure), whereas the oviparous morph lacked paravertebral markings, had less distinctly defined dorsal stripes of dark gray and light brown, and lacked belly speckling (B). They were so different, in fact, we strongly suspected they were different species (both were also distinct from the true *L. alticolor*, which was originally described from Peru). To test our hypothesis, we set out to collect lizards along the road (Ruta Provincial 307) where FML's enduring and indispensable herp collector, Omar Pagaburo (known to all as "Pagaburo"), captured the specimens that Martha had studied. Not long after we started collecting along that 16-km stretch of road (km markers 82 to 98), we discovered what Martha could not have realized from her laboratory examination of preserved specimens: the two morphs lived in distinct habitats. We found the viviparous morph exclusively in a monoculture of golden bunchgrass (*Festuca*) above 3000 m in elevation (C), and we found only the oviparous morph about 12 km northwest and 200 m lower in elevation, where the vegetation was dominated by spindly, dark-gray branched shrubs (*Parastrephia*; D). Both morphs were always encountered in the dominant vegetation of their respective regions, within which they were remarkably cryptic and nearly impossible to detect unless they moved. Ultimately, we encountered the ecotone (at km marker 88) where the two plant species co-occurred in roughly equal densities (E). Here, as might be expected, the two lizard morphs were also found in sympatry yet maintained their strict microhabitat preferences: viviparous form in the bunchgrass, oviparous form in the shrubs.

Needless to say, we were elated to have discovered a few more pieces of the puzzle, but the full picture had yet to emerge. First, the reproductively bimodal population collected by Pagaburo and documented by Martha was from km marker 95, not 88. Indeed, at km marker 95, we encountered only the oviparous morph, doubtless because there was no bunchgrass along the road at that site. Second, we couldn't explain what was driving the dramatic vegetation ecotone over such a small area and narrow elevational range.

The next morning, we arrived early to the locality of the viviparous morph. It had been cold the previous night and the humid valley some 21 km to the southeast (Tafí del Valle) was blanketed in dense fog. The *Festuca* growing along the hillsides was also covered in fog and dew. As we drove northwest along the road toward km marker 88, we realized that the fog penetrated just up to the ecotone, where it did not dissipate until early afternoon. When we returned to Tucumán, meteorological and reproductive data in Martha's dissertation confirmed that the air temperatures average about 10 °C cooler over the range of the viviparous morph relative to the oviparous morph during the peak of their respective reproductive seasons, in support of the long-held hypothesis that cold climates drive the evolution of viviparity in squamate reptiles. Furthermore, Pagaburo confirmed that he had collected both reproductive morphs of lizards at km marker

Images of the reportedly reproductively bimodal lizards (formerly recognized as *Liolaemus alticolor*) and their respective habitats along Ruta Provincial 307 in northwestern Tucumán, Argentina. A. *Liolaemus pagaburoi* (viviparous). B. *Liolaemus ramirezae* (oviparous). C. Habitat of the viviparous species (km marker 83), which is dominated by bunchgrass (*Festuca*). D. Habitat of the oviparous species (km marker 95), which is dominated by shrubs (*Parastrephia*). E. One zone of sympatry (km marker 88) of the viviparous and oviparous species; note the interdigitation of *Festuca* and *Parastrephia* along the hillsides. Photos by Robert E. Espinoza.

95 by first capturing the oviparous forms along the road and later ascending the adjacent hills, which did have *Festuca*, to collect the viviparous forms.

The puzzle pieces finally assembled into a complete image. Fernando and I wrote the first of what would be numerous papers penned by us and his future students, ultimately leading to the recognition of nine cryptic species, all of which were formerly considered geographic variants of *L. alticolor*. In that first paper (Lobo and Espinoza 1999) we also described our integrated approach to discovering the identities and ecologies of the two lizard morphs, which we named in honor of Pagaburo, *L. pagaburoi* (viviparous form; A), and Martha, *L. ramirezae* (oviparous form; B). In doing so, our investigation affirmed a widely held credo of field biologists: There is no better way to learn about the intricate lives of organisms than by studying them in nature.

Reference

Lobo, F., and R.E. Espinoza. 1999. Two new cryptic species of *Liolaemus* (Iguania: Tropiduridae) from northwestern Argentina: resolution of the purported reproductive bimodality of *Liolaemus alticolor*. *Copeia* 1999:122–140.

About the Author

Robert E. Espinoza (Bobby to most) is a professor of biology and curator of amphibians and reptiles at California State University, Northridge (https://www.espinozalab.com). He earned his BS in biology at San Diego State University and his PhD in ecology, evolution, and conservation biology at the University of Nevada, Reno, and was a Rea Postdoctoral Fellow at the Carnegie Museum of Natural History. He met his infinitely supportive spouse, Cynthia J. Hitchcock (US Geological Service field herpetologist), in grad school, and they have two teenage children, Max and Olivia, who share their parents' love for nature. His research focuses on the ecological and evolutionary physiology of amphibians and reptiles, and he relishes being in nature as much today as he did as a child.

2

THE CRAWFISH FROG'S JAW

Michael J. Lannoo

My cell phone rang. Jen Heemeyer, who was radio tracking Crawfish Frogs exiting wetlands following their spring breeding efforts, was over-the-top excited: "Frog 160 returned to her burrow, the same one she occupied all last summer." With these two facts—that Crawfish Frogs not only occupy a single crayfish burrow throughout the year but will return to that very same burrow following breeding the next year—Jen had discovered that each Crawfish Frog has a "home," a place on the planet that is uniquely theirs. Jen also understood that for this to be true, there must be some special relationship that we did not yet understand between Crawfish Frogs and the abandoned crayfish burrows they occupy.

I had recruited Jen, Vanessa Kinney, and Nate Engbrecht to be the core grad student team of researchers responsible for the Crawfish Frog project; after they graduated, I brought in Rochelle Stiles to continue the project. Our funding came through a State Wildlife Grant (SWG) administered through the US Fish and Wildlife Service. The SWG program is designed to ameliorate local-level conservation issues before they grow to become national problems requiring Endangered Species Act attention. Biologists from the Indiana Department of Natural Resources had approached me about leading the Crawfish Frog effort because these secretive frogs had been recently listed as state endangered in Indiana; I had a long, successful track record in amphibian field research; and I had the expertise after recently editing the compendium *Amphibian Declines: The Conservation Status of United States Species* (Lannoo 2005).

Crawfish Frogs (*Rana areolata*) derive their common name from their tendency to inhabit crayfish-dug burrows. Jen had shown during her previous field season that once a Crawfish Frog settled into its burrow following breeding, it rarely left—it was usually in its burrow or on a small bare spot next to its burrow, its "feeding platform." We would later discover that Crawfish Frogs will occupy the same burrow for up to five years.

As Jen tracked her frogs, she became concerned because she never saw Frog 460. While a radio signal was coming from his burrow, Jen could not determine whether 460 was alive, dead, or had shed the transmitter and moved on. We purchased a wildlife camera, set up a surveillance on his burrow, and immediately realized 460 was fine, almost always out on his feeding platform except when Jen approached. More importantly, we discovered the camera followed 460's activities with a high degree of resolution. We had a eureka moment as we realized that

A Crawfish Frog at the entrance to its primary crayfish burrow. Note the tight fit of the frog to its burrow and how the curvature of its jawline matches the curvature of the burrow, shielding the frog's vulnerable body and limbs from predators. Note also the bare "feeding platform" in front of the burrow entrance. The burrow and the feeding platform constitute a Crawfish Frog's home range during the nonbreeding season. Photo courtesy of Nathan J. Engbrecht.

for as long as Crawfish Frogs inhabited identified burrows, cameras could track their aboveground behavior. We immediately bought a dozen units and deployed them facing the burrows of Jen's telemetered animals.

Our first observation was Crawfish Frogs almost always face their burrow entrance—when they're in their burrow they look up, toward the sky; when they're on their feeding platform they face the burrow entrance. Their burrow entrance was clearly the center of their attention, if not their universe.

The burrows Jen's Crawfish Frogs inhabited were *always* dug by upland crayfish and therefore could be located hundreds of meters from wetlands. These burrows are deep (1–1.5 m)—crayfish keep digging until they reach groundwater— and narrow. After some time, crayfish will abandon these burrows and dig fresh ones; up to 40% of upland crayfish burrows across a landscape are vacant. These empty tunnels constitute the Crawfish Frog housing market.

Our second observation was astonishing and completely unexpected. We discovered that Crawfish Frogs in crayfish burrows survive snake attacks, even prolonged and determined assaults. We can think of no other North American frog species that can endure an aggressive attack by a big snake such as an adult Black Racer.

As we thought through the reasons for Crawfish Frog burrow habitation, we were assisted by Crystal Thompson's 1915 publication. She noted that Crawfish Frogs "so nearly approximate the size of the [entrance to their] holes that the rubbing of their soft bodies probably tends to smooth the walls"; "the exceptionally powerful hind limbs and the extent to which they can distend the body serving to secure them so firmly that they could be mutilated before being dislodged"; and "when alarmed the frogs do not ordinarily descend far into the burrows" (where the burrow approximates their body size). Thompson's observations, combined with our wildlife camera data, suggested that when frightened, Crawfish Frogs on their feeding platform employ a fixed-action pattern form of behavioral response comprising jumping into their burrow, turning around to face the entrance, puffing up their bodies, and lowering their heads. This scenario was confirmed when the strategy misfired.

Nate was covering for Jen. It was just after dawn, and Nate found Frog 520's signal coming from a location away from her burrow and moving much faster than a Crawfish Frog could hop. Nate caught up, pounced, and came up with an Eastern Hognose Snake sporting a suspicious midgut bulge. The terrified snake simultaneously pooped and puked. And up came 520's carcass.

It's often what doesn't fit that proves the most insightful. What didn't fit here was that 520 was missing its head, and it was obvious from the lacerated tissues that her head had been ripped off her body. Unlike sharks and crocodiles, snakes do not macerate their prey, so what had happened here? Since 520 was attacked head on, she must have been in her burrow. A frog frightened on land will flee, and therefore be attacked at the rear of its body, usually bitten on and secured by its hind legs. Frog 520 had no hind limb punctures or lacerations. Our best guess was that during the attack, 520 was in her burrow, facing the entrance. She had lowered her head, inflated her body, wedged herself against her burrow wall, and put up a great resistance. At some point, though, we think she made a mistake— she must have lifted her head. The snake got purchase on her snout, began tugging on the inflated and wedged frog, and eventually tore off her head. The decapitated frog then deflated, the snake ingested its body, and a couple hours later Nate showed up. The lesson here is that crayfish burrows and the behaviors Crawfish Frogs employ while using them allow these frogs to withstand even the most aggressive attacks by snakes, as long as they keep their head down.

At the same time Jen was conducting her telemetry study, Vanessa was monitoring drift fence arrays at Crawfish Frog breeding ponds. Vanessa noticed that the smallest breeding Crawfish Frogs were about as heavy in grams as they were long in millimeters—for example a frog might weigh 94 g and be 95 mm long or weigh 102 g and measure 100 mm. The numbers of the largest Crawfish Frogs, however, were always mismatched, with weight values consistently exceeding lengths—for example, 122 g and 110 mm. Further, these frogs often looked beat up. We couldn't have known it at the time, but a few of these animals were likely close to a decade old. For example, during our main study, which lasted from 2009 through 2016, Frog 53 bred every year. Crawfish Frogs mature at 2 or 3 years old, which would have made 53 at least 10 years old in 2016. We have similar data on other frogs. In his book *The Ecology and Behavior of Amphibians* (Wells 2007), Kent Wells points out that among frogs and toads, longevity is generally proportional to body size (large frogs live longer), but his data also suggest that the longest-lived ranids are aquatic—able to submerge to avoid predators. In a sense then, descending into crayfish burrows provides cover for Crawfish Frogs in the same way that submerging does for big aquatic ranids. For this to work, though, Crawfish Frogs had to add the behaviors of facing the burrow entrance, inflating their bodies, and lowering their heads. This is a pretty interesting narrative, and at this point we considered the story completed. Then, as they always seem to, Crawfish Frogs sprung another surprise.

In 2011, Nate led a morphological study of Crawfish Frogs. He wanted to know whether these bulky frogs, derived from more graceful ancestors, possessed morphological specializations attributable to burrow dwelling. He knew he needed to examine both adult morphology and developmental patterns among the four species in the subgenus *Nenirana*, which includes Crawfish Frogs and their closest relatives, Pickerel Frogs, Carolina Gopher Frogs, and Dusky Gopher Frogs. Nate's results showed that Pickerel Frogs have the smallest bodies, while Crawfish

Frogs, especially those in the northern portion of their range, have the largest. His data also showed that the two gopher frog species and Crawfish Frogs were thicker and had more rounded heads and shorter hind limbs than Pickerel Frogs.

The developmental shifts Nate measured were more illuminating. With growth, Pickerel Frog heads get proportionately smaller (length and width) compared with their bodies (this happens in humans, also). Similarly, in both gopher frog species head length gets smaller with growth, but head width doesn't—it remains proportional to body size. In Crawfish Frogs, head length again gets smaller with growth, but, crucially, Crawfish Frogs exhibit a marked *increase* in head width with growth. That is, contrary to most vertebrates, as a Crawfish Frog grows, its head width gets proportionally larger as its head length gets shorter. This differential growth reshapes a Crawfish Frog's jawline such that as it matures, when viewed from above, its snout approximates a semicircle.

Consistent with Nate's data, in 1953, Arthur Bragg observed that in Crawfish Frogs "the larger the size, the rounder the snout." In *The Call of the Crawfish Frog* (Lannoo and Stiles 2020), Rochelle and I suggest that a rounded snout minimizes the gap between the burrow wall and the jawline of the frog, providing fewer opportunities for predators to gain purchase. In effect, when a Crawfish Frog is in its burrow facing up, toward the opening of the burrow, and lowers its head at the approach of a predator, its rounded snout acts as a hatch that protects the softer parts of its body from a predator's teeth. This arrangement works only when a Crawfish Frog occupies a burrow with a diameter similar to the girth of its body. And it works only if a Crawfish Frog keeps its head down. Jen's intuition that a Crawfish Frog's burrow must mean much more to it than a simple hole in the ground was spot on.

As mentioned above, Crawfish Frogs are a species of conservation concern. There is a strong tendency by humans to blame victims for the crimes committed against them. This type of thinking has often underpinned informal discussions of Crawfish Frog conservation, and it goes like this: If Crawfish Frogs were not so slow, clumsy, and stupid they might muster enough wherewithal to pull themselves up by the bootstraps and save themselves from extinction; Crawfish Frogs must be held at least partially accountable for their declines.

Before our work, biologists could be excused for coming to this conclusion. Considering Crawfish Frog behaviors from the perspective of other frogs, they *are* slow. When we were processing frogs at Vanessa's drift fences, we would remove a frog from a bucket, take a measurement, set the frog down beside us, record the measurement, pick the frog up, take another measurement, set the frog down and record, and so on. This didn't work with Southern Leopard Frogs: release them and they bound off never to be seen again. But remember, unlike Southern Leopard Frogs that flee in response to predators, we now know that

Crawfish Frogs are adapted, both behaviorally and morphologically, to hold their ground when threatened. The frog Nate found in the belly of a Hognose Snake didn't die because she remained still with her body inflated and her head down. She died because she lifted her head—she moved.

References

Bragg, A.N. 1953. A study of *Rana areolata* in Oklahoma. *Wassmann Journal of Biology* 11:273–318.

Lannoo, M., ed. 2005. Amphibian declines: The conservation status of United States species. Berkeley: University of California Press.

Lannoo, M.J., and R.M. Stiles. 2020. The call of the crawfish frog. Boca Raton, FL: CRC Press, Taylor and Francis Group.

Thompson, C. 1915. Notes on the habits of *Rana areolata* Baird and Girard. Occasional Papers of the Museum of Zoology, University of Michigan, number 9. University of Michigan, Ann Arbor.

Wells, K.D. 2007. The ecology and behavior of amphibians. Chicago: University of Chicago Press.

About the Author

Michael J. Lannoo is professor of anatomy, cell biology, and physiology at Indiana University. In addition to his primary research emphasis on temperate systems, he has considerable field experience in tropical and polar regions. In 2001 Lannoo received the Parker/Gentry Award for Excellence and Innovation in Conservation Biology through the Field Museum of Natural History, Chicago, Illinois. This award honors "an outstanding individual, team or organization whose efforts are distinctive and courageous and have had a significant impact on preserving the world's natural heritage, and whose actions and approaches can serve as a model to others."

3

JOURNEY TO THE AMAZONIAN RAINFOREST

Janalee Caldwell

One of my particular interests is the reproductive behavior of frogs, especially of certain small "poison" frogs that occur in tropical regions of Central and South America. On one expedition in the late 1990s, my field party, including Laurie Vitt, Teresa C. Ávila-Pires, and Verônica R. L. Oliveira, and I worked in the state of Acre in western Brazil, near the small village of Porto Walter. This remote region of Brazil has been relatively unexplored and was known to have a high diversity of amphibians and reptiles, including poison frogs.

To reach Porto Walter, we first flew from the United States to Belém in eastern Brazil to meet our Brazilian collaborators. From there, we flew to the small town of Cruzeiro do Sul. The lack of roads in this area required us to hire a small wooden riverboat to take us to Porto Walter by way of the Rio Juruá. The boat was about 15 ft in length and had two lengthwise wooden benches in the back where the four of us sat, in addition to the bench used by the driver. Fortunately, the boat had a wooden top, although the sides were open and periodically the driving rain left us soaked.

Travel to Porto Walter required two days on the river. Initially, we traveled upstream, and because the river was flooded, as are all Amazonian rivers during the rainy season, the trip was exciting and terrifying at the same time. Our boat was constantly battered about by large branches and entire trees floating downriver. The boat motor was attached to the end of a long pole that enabled the driver to lift the motor out of the water as we glided up and over large trees. At times it seemed like the boat couldn't possibly make it over the next huge tree coming right for us, but, as we all held our breath, we made it over. When not watching for logs, we marveled at the spectacular rainforest on both sides of the river. Many of the large rivers have been heavily logged to the edge of the water, but in this remote area, large trees covered with lianas and dense vegetation were common. We spotted birds, basking caimans, and other wildlife, including several troops of noisy monkeys.

We could not travel at night because of the danger of unseen logs. Toward evening, we were beginning to get anxious because we had seen relatively few people in this remote area. Fortunately, just as it was getting dark, we spotted a small house. We met the family that lived there and explained that we were looking for a place to spend the night. They were very gracious and invited us to hang our hammocks inside their home. Throughout most of the Amazon region, the small wooden houses have large, uncovered windows, so we were lucky to have nets that covered our hammocks to keep out mosquitoes and large spiders. The next morning, we were on our way again, and after another day of wondering if we could make it over the never-ending logs, we arrived at Porto Walter toward evening.

Porto Walter was a small settlement with no vehicles or commercial buildings. As we pulled up to the dock, a few people stopped to stare at us. The mayor soon heard that strangers had arrived in town. We felt very important when the mayor himself came to greet us. Arrangements were made for us to stay in the local Catholic church, the largest building in the village. The church had many long hallways and empty rooms that gave it an eerie feeling. We were directed to a sparsely furnished room where we hung our hammocks. We questioned our guide about the long streak of blood down one of the hallways and learned that a recent murder had occurred!

The following morning, we began searching for a suitable area to set up camp for several months. We hired a local person who knew the area to help us locate a place. Ultimately, we trekked about 3 mi through swamps and fields to find undisturbed forest. The weather was hot and humid, and by noon we had finished all the water we had with us. We felt weak and exhausted when we stumbled upon a house where the family was using a hand-operated machine to squeeze sugar cane juice. Seeing the state we were in, they kindly offered us cups of juice, which tasted better than anything I have ever had to drink in my life! Eventually we located a suitable area for our camp, and the following day we returned by boat to Cruzeiro do Sul to collect our field gear and to buy enough food, mostly rice and beans, that would be our diet for the next few months.

After several days of shopping and organizing our gear, we hired the same riverboat to return to Porto Walter. During our return trip, we planned to stop for the night at the same house. We arrived in the evening and were surprised to find no one there. Eventually an older man appeared and told us the whole family had contracted malaria and was in the hospital in Cruzeiro do Sul. We were again invited to spend the night in the house, and this time we were even happier we had our mosquito-proof hammock nets!

Once we arrived in Porto Walter, we hired several local men to help transport our supplies and equipment and set up a camp at our chosen site. We began to survey the area, determine what species of amphibians and reptiles were present, and develop our individual and group research projects.

Verônica, at the time a Brazilian graduate student, shared my interest in poison frogs and was eager to participate in the research. Although it was one of her first field trips, she was enthusiastic about spending long days watching tiny frogs while not moving and fighting off the abundant, hungry mosquitoes. Her positive attitude and adventurous personality made our work much easier!

We chose to focus on a particularly common species, the Brazilian Poison Frog (*Ranitomeya vanzolinii*). Several characteristics of these frogs made it possible to study their behavior. They were active during the day, and they were black with a few bright yellow irregular spots, making it possible to sketch the spotting pattern and follow and record the behavior of the same individuals day after day. The frogs occurred in the forest understory, generally not going more than 6 ft above the forest floor. Verônica and I had no idea about

Jan Caldwell next to the Amazonian field camp. Photo courtesy of Laurie Vitt.

the amazing behaviors that we were going to observe. Our suffering with the mosquitoes in the hot, humid weather was rewarded with almost daily exciting revelations.

Unlike many species of frogs, poison frogs lay small clutches of eggs, usually fewer than 25, in leaf litter on the forest floor or in bromeliads or leaf axils in vegetation, as opposed to in ponds or other bodies of water. Once the eggs hatch into tadpoles, the male parent (but sometimes the female, depending on the species) carries the tadpoles on its back either singly or in a group to temporary pools, streams, or a water-holding plant such as a bromeliad, where the tadpoles complete their development and transform into frogs.

Initially, we found that we could locate the same male frogs in the same general areas every day, indicating that they maintained individual territories. We often observed pairs of males alternating calls, announcing the boundaries of their territories. As we got to know individual frogs from their yellow spots, we were surprised to find that the same males and females were always found together in the same territories.

It was easy to distinguish males from females. Females were heavier bodied because they were frequently gravid with developing eggs. Although males called nearly every day, their calls became more intense and more frequent when a female appeared. Further, when a female appeared in a male's territory, we noticed that the female followed the male as he moved around.

At times, we found ourselves laughing at the frogs' antics. As scientists, we avoid anthropomorphic terms when describing behavior in animals, but occasionally we couldn't resist viewing their behavior in human terms. One male with a female following him took a huge leap from a large leaf to another leaf about 3 ft below. Instead of landing on the leaf, he bounced off the leaf high into the air and landed on the ground. He sat there for a long time, not moving, and appeared to us as if he wasn't sure what had just happened. He eventually recovered and climbed the nearest tree. On another occasion, a fat, gravid female was following a male. The male jumped off a big leaf and landed on another leaf about 4 ft below. The female moved to the edge of the leaf and looked over. From our human point of view, she seemed to be thinking "no way!" as she backed off. Meanwhile, the male called frantically from below, trying to convince her to jump. After coming to the edge and backing off three more times, she finally got the courage to jump and landed safely.

Ultimately, after watching the interactions between males and females, we discovered that males were not leading females around randomly but were leading them to small holes in a sapling tree or a woody vine. Often the female, but not the male, would enter the hole, remain for a short period of time, then leave and not reappear for several days.

At first, these observations were puzzling to us. We had also seen several males, each with a single tadpole on its back, enter holes and exit without the tadpoles. We investigated the holes with a flashlight and discovered that many of them held a small amount of water and contained a tadpole. In some holes, two or three small eggs were attached to the side of the hole above the waterline.

Because the holes in which these eggs and tadpoles were deposited were very small, behavior of the frogs while in the holes was difficult to watch. Therefore, we placed several small plastic containers filled with water on a log a foot or so above the forest floor. After a few days, two tadpoles appeared in one of the pans. While watching the tadpoles, we observed a female poison frog approach and back into the water. As she did, the tadpoles approached her and began to nibble at her back legs. This activity stimulated the female to release an egg into the water. The tadpoles immediately attacked the egg, breaking through the outer membranes and consuming the yolk. Verônica and I were amazed at this observation and immediately began piecing together our observations.

We ultimately realized that males and females of these tiny frogs were cooperating to raise their young. It was truly a surprising revelation at the time because biparental care in frogs had not been described. (In recent years, a few more species of poison frogs have been found to have this behavior.) The male's role, in addition to fertilizing the female's clutch of two to three eggs, included transporting the tadpoles individually to a small vine or tree hole and later guiding the female to the holes so that she could provide nutritive eggs for the tadpoles. The small opening of the holes did not allow sunlight to enter, so the females' eggs were the only food available to the tadpoles. A female frog was capable of producing eggs every four days; thus, the tadpoles were periodically fed until they metamorphosed into froglets.

Many questions remain unexplained about the behavior of these frogs. For example, as in birds, could a male be cooperating to raise young with more than one female at a time? We were shocked near the end of our time in the field when we discovered one of the males we were following was leading an unknown (to us) female to a small tree hole. We had seen this male previously leading a different female to a hole with a tadpole. Could this male have two clutches he was caring for with different females? We were unable to follow the frogs for more than one season, so it is unknown whether the same males and females bond to care for offspring in successive seasons. Indeed, we do not even know whether these tiny frogs have a lifespan of more than one year. Like almost all research, many additional questions arose as we continued to learn about the behavior of these frogs.

After several months of living in the forest, it was time to return to civilization. We were grateful to the people we met in Porto Walter for their kindness in helping us with so many different aspects of our trip. We visited many of them on our last day in the village to say farewell and express our gratitude. The trip back, again by riverboat, required one day, rather than two, because it was much faster to travel downstream. We were happy to see that fewer logs were floating down the river as the end of the rainy season approached!

About the Author

Janalee Caldwell received a BS in zoology at Oklahoma State University and a PhD in systematics and ecology at the University of Kansas. At the University of Oklahoma, she was a professor of biology and curator of amphibians at the Oklahoma Museum of Natural History. She spent most of her research career studying the behavior and systematics of frogs in Brazil. She has authored numerous scientific publications and is coauthor of a textbook entitled *Herpetology: An Introductory Biology of Amphibians and Reptiles*. In 2004, she was awarded the University of Oklahoma Regents' Award for Superior Accomplishment in Research and Creative Activity and is currently professor emeritus and curator emeritus at the University of Oklahoma.

4

A RAINY EVENING IN THE PANTANAL

Cynthia P. A. Prado

One of the wonders of being a field biologist is witnessing events no one has ever seen. In late spring of 2000, something special happened during my fieldwork in the Brazilian Pantanal, an event that would guide my academic career.

With approximately 210,000 km², the Pantanal is one of the world's largest floodplains. It is located in central South America, mostly in Brazil but extending into Bolivia and Paraguay. The flat and open landscape, dominated by native grasses and some patches of forest, resembles a savanna with a patchwork of rivers and seasonal ponds. It is a place full of life, and of extreme contrasts. Temperature often exceeds 40 °C in spring/summer, but during the winter it may suddenly drop to zero or below. Floods are seasonal and may last four to five months. At the peak of the dry season, in September or October, the landscape becomes xeric, green leaves turn shades of gray and brown, and depending on the year, most of the ponds dry out, the rivers barely flow, and animals perish. But this cycle of floods and drought sustains the Pantanal's rich biodiversity and enormous animal abundance.

I fell in love with the Pantanal and frogs when I was an undergraduate student at the Federal University of Mato Grosso do Sul. During my first year in the biology course, one of our professors organized a trip to study the embryonic development of amphibians in the field. The Pantanal was completely flooded, and our bus could not reach the university's research station. Instead, we rode in boats to the station. I was very excited with that adventure and the opportunity to see all the caimans, capybaras, and birds up close. Of course, not everything was so pleasant; it was extremely hot and humid, and there were tons of mosquitoes around us. At night, we finally went to the ponds to collect frog clutches with our professor and a veteran colleague who knew the species. I was really impressed by her ability to listen and identify the frog species only by their calls, and I thought to myself: "I want to be like her!" I can say that I was successful.

Between 1992 and 1999 I conducted fieldwork around the research station at the margins of the Miranda River, about 370 km from the main campus. During six consecutive years, I visited the field at least once a month to collect data for my undergraduate and master's work. In 2000 I began my doctorate and moved to São Paulo state. I decided to continue my study of the frog reproductive strategies in the Pantanal, as I had collected enough data for most of the species except the frogs that reproduce only following heavy rains and then disappear. Contrary to what many people think, rainfall is low in the Pantanal and heavy rains are rare. The floods are mainly due to rains in the surrounding highlands; water of the tributaries flows down to the plain, causing the main Pantanal rivers to overflow. I needed to live in the Pantanal. For two rainy seasons, from October 2000 to March 2001, and the same period the following year, my home was the Pantanal research station. Those were the happiest days of my life. No internet, no cell phone signal, no TV or radio. Most of the time I had little contact with people. The mornings were welcomed by the sound of many different birds and the sun rising over the river, with Jabirus flying to cross to the other side.

I had a borrowed laptop, and I took along two boxes—one full of scientific papers and another with my favorite CDs and a CD player. My doctoral advisor, Célio Haddad, lent me a Canon film camera and a Uher sound recorder. I arrived at the Pantanal research station at the peak of the dry season, in early October 2000. Everything was dry and extremely hot. There were no ponds nearby, and the river level was very low. After a month of searching the sky for clouds, finally a heavy rain arrived in mid-November. It rained the whole night, and a pond formed beneath the research station, which had been constructed on stilts as protection from floods. The next day remained cloudy and the rain turned to drizzle. In the afternoon, I heard a frog call I had never heard before, even after all the years collecting in this area. I changed into my field clothes and rubber boots, grabbed my field equipment, and began exploring the recently formed pond.

Suddenly, everything changed after the rain. There were at least 15 different frog species calling at the same time in that recently formed pond. In the middle of that explosion, I listened to that unknown low-frequency call again. After searching for a while, I finally found the frog. I was surprised that the frogs calling were males of the Miranda's White-Lipped Frog (*Leptodactylus macrosternum*, formerly *L. chaquensis*), a very common species in that region. Males can attain 8 cm in length; they have hypertrophied arms and spines in their thumbs that are used in male–male combat. I already knew that the species was a foam-nesting breeder and that females cared for eggs and tadpole schools, easily observed in spring/summer. However, during my previous visits to the field, I had not observed a breeding event because the species is very explosive, reproducing only for one or two days. After many years of fieldwork, I was listening to their calls and observing their spawning behavior for the first time. Amazing!

It was midafternoon, on 13 November 2000, still cloudy and drizzling, when I observed a pair of *L. macrosternum* spawning, but they jumped out from the foam and hid in the vegetation when I approached. At dusk, I returned to the same site, and there were two males calling and fighting in the middle of the foam nest previously initiated. An hour later, there were four males calling and fighting. The resident male was large, and the other males were visibly smaller in size. I was by myself and so excited that for a moment I could not believe I was observing this behavior. I did not know what to do at first, as I did not have a video recorder with me, so I started taking pictures and tried to record their aggressive calls. The frogs continued wrestling, and the resident male eventually clasped another male with the forelimbs, both coming out from the foam chest-by-chest pressed together. Additional males showed up, and when there were eight males wrestling, a hidden female jumped into the nest, and the resident male grasped

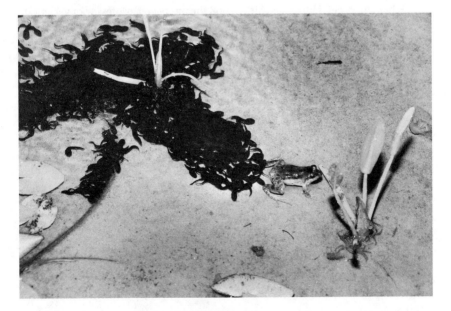

Another fascinating frog from the Pantanal is the Pointedbelly Frog (*Leptodactylus podicipinus*). Here, a female is caring for a school of tadpoles in a pond during flood season in the Pantanal, Brazil. Photo courtesy of Harry Greene.

her and immediately they started spawning. At the same time the other seven males began to kick the foam nest with their legs in synchrony with the amplectant male. At that exact moment, the film of my camera finished. Fortunately, based on some of the pictures I took, a technician from our department could draw the multimale spawning behavior, and I was able to publish this observation, which my advisor, Célio, and I interpreted as an alternative reproductive behavior, with sneaker males trying to fertilize some of the eggs released.

I remained at the research station until March, going to the field almost every night, and also the following summer, but I never observed that behavior again. At the time, although there were some reports of simultaneous polyandry for a dozen frog species, this was the first record of this behavior for a South American frog. I am very happy I could witness that behavioral event on that rainy evening in the magical Pantanal floodplain.

About the Author

Cynthia P. A. Prado graduated in biology from the Federal University of Mato Grosso do Sul and earned a PhD in zoology from the São Paulo State University (Universidade Estadual Paulista, UNESP). During her postdoc she was a visiting researcher at Cornell University. Since 2008 she has been a professor of vertebrate biology at UNESP, São Paulo state, Brazil. In 2018 she was one of 13 Brazilian women scientists to be honored with the Bertha Lutz Prize, awarded by the Amphibian Specialist Group/ International Union for Conservation of Nature, in recognition for her contribution to amphibian conservation. Her research focuses mainly on ecology, evolution, and reproductive behavior of frogs.

5

TRACKING TURTLES

Jacqueline Litzgus

It is the middle of an early summer night, I am by myself under the majestic canopy of North America's last old-growth cypress swamp, sweating as I stand stock-still behind a tupelo tree, surrounded by several species of venomous snakes and alligators and other unknown bitey things.

I wonder, "Did I remember to tell anyone I'd be out here tonight? Is my cell phone fully charged? How far am I from my truck?"

Wait, I've just had a déjà vu. . . . I've been in a situation like this before, but swap the swamp for the ancient and smooth rock barrens of the Canadian Shield and the alligators for black bears, add lightning, and insert a canoe as my mode of transportation.

I wondered then, "Can I get back to my canoe from here, and how close is that bear I hear gruffling?"

These are the memorable moments of radio tracking gravid female Spotted Turtles (*Clemmys guttata*) to find their nests in South Carolina and Ontario during my doctoral and master's fieldwork, respectively.

The Spotted Turtle is a globally endangered species whose populations are declining because of habitat loss and modification, road mortality, and illegal collection for the pet trade. They are really pretty, charismatic little turtles—their black shells adorned with a constellation of yellow polka dots—and so they are highly desirable as pets. Conservation of the species requires describing and protecting the wild habitats they need to complete their life cycle, and this requires fieldwork. Protection also requires keeping secret the locations of field research sites (https:// theconversation.com/the-illegal-turtle-trade-why-i-keep-secrets-85805).

I became interested in Spotted Turtles while working toward my undergraduate degree. I taught biology for a summer job at a canoe-tripping camp for kids in eastern Georgian Bay, Ontario, Canada. At the camp, I found old data sheets reporting information about individual Spotted Turtles captured around the camp by campers and staff; it was like opening a map to a treasure chest! And then I found those exact same turtles when I took kids out on programs. It was like finding the actual treasure! Previous camp staff had individually marked all the captured turtles by filing different combinations of small triangular notches around the outside edge of their shells, so I knew who was who. Finding these turtles made me ask all kinds of questions about them, and I turned that curiosity into my master's thesis research studying the spatial ecology and life history of northern Spotted Turtles.

Like all new graduate students, I hit the scientific literature to see what we already knew about Spotted Turtles, with specific attention on their reproductive behavior. All the previous work was focused on populations in habitats very different from the rock barrens and shrub swamps of Georgian Bay, so I was going in blind. From the literature, I *did* know that it was unlikely they would nest during the daytime, so I needed to locate them at night. I outfitted gravid females with radio transmitters and tracked their movements to nesting sites. On that one particular night referenced above, I made the surprising discovery that the turtles nested on rock outcrops. Yes, I said rock.

I was tracking female ID number 2 (later named "Mom" for reasons that will shortly become obvious). She was gravid and ready to pop; she was full up to her armpits with shelled eggs, white elliptical jewels of potential. During the day I had located her in a wetland at the base of a rock outcrop. That night, her radio transmitter signal was coming from up on top of the rock. I thought, "Oh no, she's been eaten by a predator who dragged her up onto land!" I followed the signal, thinking I'd find her carcass, and then stood befuddled on the top of the rock barren, cursing the rock itself and the rain for making the telemetry signal bounce around so that I couldn't locate her. In frustration and defeat, I dropped my chin down to my chest, and lo and behold, between my feet, revealed in the luminescence of a lightning flash, the lichen wiggled. There she was! Wearing the lichen on her shell like camo on an army helmet as she excavated a nest chamber in a shallow pocket of soil in a crack on the rock. Yes, I said rock.

Now I knew what to look for. All the Spotted Turtles I tracked during my master's fieldwork laid their eggs on rock outcrops. And subsequent fieldwork revealed that all the freshwater turtles we and others studied in Georgian Bay also made their nests on the rock barrens. These heat-holding rocks, with their thermal inertia, are egg-incubating hotspots, ideal for embryogenesis in a place with short and cool summers.

So then I wanted to know where Spotted Turtles in warmer places without rock barrens would lay their eggs, which brings me back to the cypress swamp full of cottonmouths and other venomous snakes in South Carolina. When I was standing behind that tupelo tree, I was quietly watching a gravid female Spotted Turtle. She was standing on the top of a large fallen cypress log, and I was again befuddled. I wrote in my field journal, "What is she doing? Moonlight basking?" Well, it turns out she was nesting. She laid her eggs in the rotten wood and moss on top of the log. Another surprise! Most of the other female Spotted Turtles tracked during my doctoral field research also nested in rotten woody debris within the swamp forest, but not all of them were on the tops of logs like this moonlight basker. My guess is that in warm places, turtles lay their eggs in shady spots to prevent the eggs from overcooking, which could cause deformities or even embryo mortality. These canopied swamp forests are egg-incubating cooling spots, ideal for embryogenesis in a place where summers are long and hot.

I have been working on Spotted Turtle ecology and conservation for three decades now and have had the opportunity to visit other field research sites in Canada and the United States to survey for Spotted Turtles. I am amazed by the diversity of habitats in which the species lives, despite its at-risk status. Over those years, I have observed changes and declines in turtle populations, and my life has also changed, including having a family. My daughter has accompanied me to some of these field research sites.

Spotted Turtles are typically one of the first turtle species to emerge in spring, which means surveys in northern locations often happen when there is still ice and snow skirting the hibernation wetlands. On one memorable early spring survey, my daughter, who was only seven years old at the time, joined the research team, outfitted in two pairs of my wool socks and my hip waders, which were chest waders on her small frame. She valiantly attempted to find turtles, not getting discouraged by her lack of captures. She was content eating snacks of Goldfish crackers and pepperettes during the pauses when we sat on the rock outcrop to measure and weigh the captured turtles—that is, until a few hours into the fieldwork day when she got a "soaker," a Canadian term for when you take water into your boots or waders. The water was only 4 °C. No amount of wool can keep little feet inside hip waders warm for long at that temperature. So as she stood shivering in sopping stocking feet on the rock, dumping the frigid water from her waders, she very calmly said, "Mommy, I think I'm ready to go home now." And so we did, and she slept in the backseat of the car for the whole drive home. Sharing these field research experiences with my daughter is fulfilling on a personal level but also on a broader scale because it helps curb the contemporary childhood problem of "nature deficit disorder."

Jackie Litzgus's daughter holding a Spotted Turtle captured during early spring surveys in eastern Georgian Bay, Ontario, prior to getting a "soaker" in the frigid waters—an event that ended a productive day that merged field research, family life, and building empathy for nature. Photo by Jackie Litzgus.

I learned from radio tracking Spotted Turtles in two geographically distant places that fieldwork is hard, even grueling, but so rewarding. To be that intimate with, and embedded in, the natural world is an unparalleled opportunity and experience for building empathy for the slimy and scaly creatures with whom we share this planet. It's hard to love something you don't understand. So get outside, learn, and fall in love.

About the Author

Jacqueline Litzgus grew up catching snakes, turtles, and toads in the forests and creeks near her house and had the rare privilege to turn her childhood fascination into a career. Now as a professor, she has the opportunity to share her passion for these animals and their conservation with students in the classroom and in the field. Jackie has been a professor at Laurentian University in Ontario, Canada, since 2004. She completed her BSc and MSc at the University of Guelph and her PhD at the University of South Carolina. Jackie was honored to receive the 2020 Distinguished Herpetologist Award from The Herpetologists' League. Jackie is still working at her field site in Georgian Bay, monitoring those Spotted Turtles that surprisingly nest on rocks.

6

FINDING THE FROG THAT SINGS LIKE A BIRD

Jodi Rowley

I stood with a group of colleagues and students at a small village at the base of Pu Hoat mountain in Vietnam. It was the monsoon season, a time when most biologists head *out* of the drenched forests, but we were about to venture in. Staring up at the sheer mountain face, surrounded by dense evergreen forest, we were about to begin our search for a tiny, mysterious, turquoise green frog. We knew we were looking for the proverbial needle in a haystack, a miniscule frog barely half an inch in body size—an almost transparent greenish frog against a backdrop of lush green vegetation, in the vast mountainous forests of Pu Hoat Nature Reserve in northwest Nghe An Province, Vietnam.

Our journey began about nine months earlier, and in a very different setting. Sitting in front of my computer in my office at the Australian Museum in Sydney, I opened an email from my colleague Vinh. Vinh had attached photos of frogs he'd seen on a recent field trip. I'd been working on the amphibians of Vietnam with Vinh and colleagues for years, and it's always exciting to see photos of the frogs and salamanders that call the mountains of Vietnam home. I flicked through the photos of some spectacular but familiar frogs— tiny brown horned frogs calling from the leaf litter, chubby little tree frogs perched on leaves, and mottled cascade frogs with enormous toe pads clinging to waterfalls. And then, there was a photo of a frog I hadn't seen before! A tiny, almost transparent turquoise frog with flashes of yellow along its side and curiously black soles on its feet. This frog was completely unfamiliar to me, and even more intriguing, I knew of no known frogs in the region that resembled this tiny gem.

It was that single photo that brought my colleagues and me to this far-flung montane village at the edge of the forest at the hottest and rainiest time of year. We were in search of a mystery frog. However, before we could even think about finding the frog, we had to get up and into the forest. In actuality, it wasn't very far in terms of distance from the village where we stood to where we hoped we might find the frog. However, the forest was dense, the paths were muddy and tangled with vines and fallen vegetation, and the route up the mountain was extremely steep. One particular section (which I labeled in my GPS as a phrase that I shouldn't share) was an almost vertical slippery slide of mud and leaf litter. Lacking agility, I fell face-first several times and had to scramble up most of the steepest sections on all fours. Despite the challenging terrain, we climbed up and up, pausing occasionally to catch our breath before slipping and sliding up the mountain. The farther away from signs of the village, the closer we hoped we were to the mystery frog.

After several hours of sliding *up* the mountain, we reached about 1500 m above sea level. As we hiked, the forest had become incredibly dense and the track less and less obvious until we lost the track and the way completely. We stopped and scouted around until we found a small, clear stream. We'd all run out of water by then, so we gratefully drank from the stream, then sat on nearby rocks and ate dry biscuits and sticky rice before hiking up to a larger stream. Huge, round boulders flanked this fast-flowing stream and lush vegetation filled every nook and cranny. It was here that we made base camp, some of us in hammocks with tarps overhead, others simply laying tarps under the shelter of a huge streamside boulder.

After a meal cooked on a campfire, we donned head lamps and headed out in search of frogs, as darkness, and rain, fell. One of my favorite things in the world is to hear the sounds of the forest erupt as it gets dark. Most frog species call only at night, and as the sun sets the calls of male frogs begin to fill the air. In a forest in which the frogs haven't been scientifically studied, this is particularly exciting. Amongst the honks, creaks, and whistles of frogs that were familiar to me were some strange and unusual calls. Some of these were likely to be frogs that I'd just never heard myself, but others were likely frog species that were unknown to science. One might even be the tiny green frog we were so keen to find. These strange calls lured us out into the night.

Walking alongside the stream, we heard frogs calling intermittently from within boulders in the streambed and from up stream banks. Seeing the frogs responsible for these calls, however, proved a challenge. The calling frogs were easily spooked and stopped calling as we got near; even if we thought we knew where they were, they seemed impossible to find in the thick vegetation. Indeed, the vegetation around us was some of the most dense that I'd ever seen—perhaps no surprise as the forest was misty most of the day and night, perfect conditions for plants to cling to every surface.

After a few hours of searching, we'd found only a handful of frogs and recorded their unique calls. While we recorded important information on these frogs, we didn't find the frog we were really after. We returned to camp in the early hours, by which time it had gotten rather cold. The cool temperatures were exacerbated by the fact that we were all completely saturated from our night hiking through wet vegetation and streams. The water from the persistent mist clung to our bedding—and us—making for a soggy night all around.

After drying out slightly, and a day scouting around our main camp for sites to survey at night, we ventured farther from our camp the following afternoon. We found a slow-flowing trickle of a stream, and as darkness fell, we set about recording the cricket-like calls of some unusually large leaf-litter frogs. Crouching in the dark, we held microphones toward the calling frogs, turned our lights off, and hoped that the frogs would continue calling. Each recording provides vital information on the identity of the frog, including whether the frogs are a species known or unknown to science.

Suddenly, we heard a strange bird-like twittering from a boggy area to the side of the stream. We rushed over to the source of the noise and there, right in front of me, perched on a leaf, was a tiny, shimmering, turquoise frog!

Quang's Tree Frog (*Gracixalus quangi*), Pu Hoat Nature Reserve, Vietnam. Photo by Jodi Rowley.

The call of this frog species was like nothing I'd ever heard before. Usually, a frog makes a repeated call ("ribbit, ribbit, ribbit"). Sometimes they will have a little more complex call ("reeek-pip-pip"). But the call of this frog species was bizarre. No two calls were the same; there was a mix of high-pitched chirps, whistles, and clicks. The overall effect was like being surrounded by dozens of tiny, excited, high-pitched birds.

I was in absolute heaven! For the next few hours I stayed in this little grotto of turquoise frogs, trying to get clear recordings of as many calling males as possible. Whilst walking around the site trying to spot the frogs, I came across two clutches of eggs, each attached firmly to the tips of leaves. Only in the wettest of forests can frogs risk laying eggs outside water, where they could dry out if the humidity dropped. The eggs resembled small clusters of pearls, clinging to the tips of leaves, causing the leaves to droop at the tips. Tadpoles of this species must face quite a shock as they break free from the eggs and then drop into the water below to continue their development.

We returned to camp elated, and for the next week we continued documenting the frogs of this mountain before returning to our more office-bound lives. From our fieldwork, and follow-up work in the lab, our team was able to describe these frogs formally as new to science, naming them Quang's Tree Frog (*Gracixalus quangi*), in honor of our colleague. We also scientifically documented the call of this species, and in doing so realized that Quang's Tree Frog had perhaps the most complex and variable frog call known. As a result, the tiny turquoise frog that lays its eggs on the tips of leaves has been coined "loài ếch có tiếng kêu như tiếng chim hót," or "the frog that sings like a bird."

About the Author

Jodi Rowley is a conservation biologist obsessed with amphibians. Based at the Australian Museum and the University of New South Wales in Sydney, Australia, her research seeks to uncover and document biodiversity and inform conservation decisions. Jodi obtained a PhD supervised by Ross Alford at James Cook University. She has led many expeditions in search of amphibians in Australia and Southeast Asia and has codiscovered more than 30 frog species new to science, including the Vampire Flying Frog. She is the lead scientist of FrogID, a national citizen science project developed by the Australian Museum that has collected almost a million records of frogs across Australia since 2017. Jodi is a fellow of the Royal Zoological Society of New South Wales.

7

BORNEO'S TADPOLE HEAVEN

Indraneil Das

I remember the walk clearly. It was August 2002, and my friend and colleague from Germany, Alex (now Professor Haas), then with the University of Jena, and I were slowly trudging up the summit trail at Kubah National Park. Night was just falling, and we had set out to find a good spot to find frogs, particularly their larval stages.

Having arrived in 1998, I was a new staff member at the brand-new Universiti Malaysia Sarawak, located near the city of Kuching, in Sarawak State, on the island of Borneo. My previous experience on the island had been in adjacent Brunei Darussalam, and my tadpole samples from the time were kept aside as I grappled to identify the often-complex (adult) frog fauna of the rainforests of that country. Now, with the arrival of a self-confessed "tadpole guy," I was more hopeful of arriving at precise determinations. Alex, the son of a watchmaker, has both the eye and patience for small details, just what's needed to study tadpoles.

Approaching what looked like the top of the ridge, we (at the time a fairly physically unfit duo) were relieved to hear frog calls. And not just one or two, as in most Bornean forests, but a crescendo of a multitude of species, some of them unknown to our fairly well-trained ears!

As we gravitated to the left of the trail, a few yards along a muddy bank led us to an incredible sight—a largish pond, the result of excavation for an arboretum and other horticultural activities, on whose edges and on overhanging plants and adjacent trees sat a multitude of species, some clearly recognizable, others not. Areas in the water were dense with tadpoles, feeding or coming to the surface en masse to take hasty gulps of air, avoiding the equally numerous dragonfly naiads out to eat them. Given that standing bodies of water of this size are a rarity on the island (and most species breed in far smaller water bodies, such as temporary puddles and oxbow lakes and depressions created when large trees are uprooted), we realized that here was a great site for future research.

Some words on the pond itself are in order. This unimpressive-looking, often muddy depression is irregularly oval, 25–40 m at its greatest length and 8–12 m at its maximum width. Depending on the season, it is 1.5–2 m deep and edged with a low earth embankment that offers a perfect habitat for a variety of anuran amphibians.

Kubah's frog pond sits at the foot of the Gunung Serapi sandstone massif, within the Matang Range, from where we have recorded 70 species of amphibians. Here, in a relatively small area, are as many as a dozen species that breed, some simultaneously. Studies by our Malaysian and European students, Alex, and me have produced several papers and theses on life histories, new records, and taxonomic descriptions of species; some of the studies continue to this day. Work at the pond and its surrounding areas has been done, for instance, on the female advertisement calls of the Smooth Guardian Frog (*Limnonectes palavanensis*); the breeding biology of what must be one of the smallest rhacophorids, White-Eared Tree Frogs (*Feihyla kajau*); and multiple-male amplexus in the Harlequin Tree Frog (*Rhacophorus pardalis*).

It is here that we found the region's smallest frog (breeding males less than 13 mm in snout–vent length), the Matang Narrow-Mouthed Frog (*Microhyla nepenthicola*), whose tadpoles develop exclusively in the pitchers of a particular species of *Nepenthes* (pitcher plant), and the Red Hot Chili Pepper Frog (*Glyphoglossus capsus*), an attractive frog that is olive dorsally and blood red on the throat region. A visiting sound enthusiast from Australia made recordings of the pond by night, highlighting the Mahogany Frog (*Abovarana luctuosa*), and the clip won a competition for the best sound in the world, organized by the lifestyle website, BeautifulNow.is!

Overall, a visit to the pond is never disappointing and is a great opportunity to see many large, charismatic species among the frog fauna of Borneo that are seldom seen elsewhere, such as the File-Eared Tree Frog (*Polypedates otilophus*), the Bornean Tree Frog (*Rhacophorus borneensis*), and the celebrated Wallace's Tree Frog (*R. nigropalmatus*). In addition, not far from the pond itself, we have observed both species of caecilians known from the Matang Range, the Broad-Striped Caecilian (*Ichthyophis asplenius*) and the Biangular Caecilian (*I. biangularis*). However, it is the knowledge generated on the larval stages of amphibians here in the remote pond and its vicinity, in the middle of nowhere, that we find most fascinating.

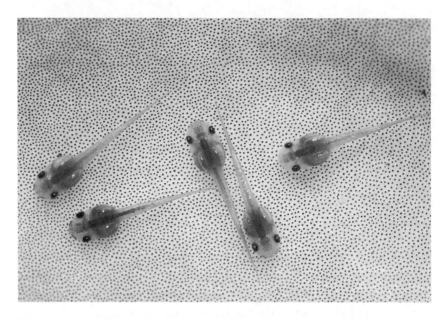

The tadpoles of Matang Narrow-Mouthed Frogs (*Microhyla nepenthicola*)
develop exclusively inside the pitchers of *Nepenthes ampullaria*. Photo by
Indraneil Das.

One species encountered at the edge of Kubah's frog pond does not actually
use these waters to breed and instead shows unusual parental care. The Smooth
Guardian Frog lays and fertilizes eggs in the moist leaf litter, dozens of meters
from the nearest stream. It is the task of the males to attend to the eggs. Upon
hatching, the tadpoles climb onto the back of the father, who carries them to the
stream, where they develop and transform into little frogs. Farther away from
the pond are several other species of frogs that have even less use for ponds and
rivers. These include the bush frogs (*Philautus*) and the dwarf toads (*Pelophryne*),
both completing development inside water-filled plant structures, from pitchers
of the local species of pitcher plants (*Nepenthes*) to tree cavities. Both groups
are endotrophic, obtaining all the energy required for growth and transforma-
tion into little frogs from the yolk supplied by the mother. Our favorite frog, the
Matang Narrow-Mouthed Frog, also shows such adaptations. Its tadpoles, about
11 mm in maximum length, inhabit the pitcher plant *Nepenthes ampullaria*. Also
endotrophic, it does not have an oral aperture and has a remarkably short gut,
with only three "bends." The tadpoles turn into little frogs within a fortnight.

Just behind the frog pond is a shallow, flat stream that flows over sand, draining into the torrential (and oddly named) Sungei Bawang (= Garlic River), where
herpetologists have broken more than one bone trying to study the tadpoles that

live in this dangerous (to humans) environment. In fact, the current is so fast that it's a miracle any tadpole should occur here in the first place. Two groups of amphibians regularly utilize such environments: the slender stream toads (*Ansonia*) and the Bornean torrent frogs (*Meristogenys*). The tadpoles of both species show peculiar adaptations, including oral suckers and in some even belly suckers, to cling to smooth surfaces of rocks. Such is the efficiency of the suckers that it is sometimes a challenge for field investigators to pry the tadpoles loose from such places using bare hands (and nails)! To understand how suction is achieved in *Meristogenys*, we studied the abdominal and oral suckers by means of three-dimensional reconstructions of serial histological sections. Our observations show that the suckers are composed of muscles and ligaments attached to the skeletal structures, and some muscles attach directly to soft tissue, explaining their significant strength.

Also nearby, but in the quiet side pools or bends of such wild mountain rivers, are gentle habitats—deep pockets of rock, filled with dense leaf litter. These areas create yet another unusual condition: a world of near darkness, anoxic conditions, and low temperatures. In such microhabitats develop the larvae of yet another specialized frog, the so-called foot-flagging frogs (*Staurois*). The adults, which are chiefly diurnal, raise their legs to communicate with each other against the thunderous background of the cascading waterfall. Their tadpoles are eel-like, and as might be expected of the habitat occupied, they lack pigments. The tadpoles appear in shades of red, pink, and orange, and they panic when exposed to the light of day.

The tadpoles of the puddle or seep frogs (*Occidozyga*) occupy shallow depressions and can swallow entire insect larvae, a feat for an amphibian larva. Our studies on its morphology show adaptations such as a relatively small mouth and tunnel-like arrangement of structures that form the buccal cavity, suggestive of suction feeding that generates strong negative pressure, much like a vacuum cleaner. Tadpoles of the genus *Leptobrachella*, referred to as slender frogs, live in shallow but highly oxygenated waters flowing over fine rocky sediments. They can merrily enter and exit such microhabitats, flexing their heads and necks in a way mammals do and other tadpoles can only dream of. Again, our detailed morphological studies, looking at their cranial bones and associated muscles, show how this is achieved. The mobility of the head is attributed to the freeing of the notochord from the base of the cranium, as well as the articulation of the foramen magnum floor with the atlas, conditions unknown in other frogs. Further, certain trunk muscles reach far forward onto the skull. Finally, with a total of 35 vertebrae, these tadpoles are indeed very long! Those who find the Bornean Horned Frog (*Pelobatrachus nasutus*) bizarre would find the biology of their larval stages equally so. These tadpoles have an upward-directed oral disc that spreads under the surface film of the water, sucking in microscopic particles of food (including algal material, pollen, and perhaps also bacteria) by means of the branchial pump.

Frog biology is indeed engrossing, be it for researchers or the general public. And the natural history of their larval stages—with its quirks and outright strangeness—must be one of the most diverse in the entire vertebrate world. In the intervening years, the fame of the frog pond has spread far and wide, and in recent years, few but the wettest nights have deterred ecotourists and photographers to the pond. Many travel agencies specializing in wildlife tours in town even sell frog tours to the pond. An event we at Universiti Malaysia Sarawak conduct annually (since 2012) with partners, such as the Sarawak Forestry Corporation, is the International Bornean Frog Race. Holding the race, a photographic competition with a variety of other programs and sometimes other parallel events at Kubah centered around the frog pond, has been a great success. The pond provides plenty of frogs for participants to photograph, and the event highlights the importance of amphibians through the media. We cap entry to 100 persons to reduce impact on the environment. Up to 17 countries have been represented at a single year's event, making it a truly international affair.

The fame of the frog pond at Kubah National Park, admirably managed by the Sarawak Forestry Corporation and the staff at Kubah, continues to draw researchers and has helped bring a sense of pride to the people of Sarawak regarding their natural heritage, including frogs, which are often deemed lesser than hornbills and orangutans. We are happy to be part of that journey.

About the Author

Indraneil Das is professor with the Institute of Biodiversity and Environmental Conservation, Universiti Malaysia Sarawak. His early herpetological work was conducted while at the Centre for Herpetology/Centre for Island Ecology, based at the Madras Crocodile Bank, India. Das received his doctoral degree from the University of Oxford and was a postdoctoral fellow at Universiti Brunei Darussalam and a Fulbright Fellow at Harvard University. His current research interests include systematic zoology, community ecology, conservation biology, biogeography, and the history of natural history. Das's fieldwork has taken him to Asia, South Africa, Europe, and the United States. Awards include a gold medal and Honorary Fellowship of the Zoological Society of Eastern India; Distinguished Herpetologist Award of The Herpetologists' League; and Honorary Foreign Member, American Society of Ichthyologists and Herpetologists.

8

HOW THE BOG FROG GOT ITS NAME

Paul Moler

The Pine Barrens Treefrog (*Hyla andersonii*) was first recorded in Florida in 1970, having been previously known only from New Jersey, the Carolinas, and a single report from eastern Georgia (Christman 1970). Early surveys located only 11 breeding sites in Florida, and 4 of those had been destroyed by development when, in 1977, the Florida population of the Pine Barrens Treefrog was determined by the US Fish and Wildlife Service (USFWS) to be Endangered under provisions of the Endangered Species Act (USFWS 1977).

I began working as a herpetologist for the Florida Game and Fresh Water Fish Commission in December 1977. One of the first projects I began was a survey for the recently federally listed Pine Barrens Treefrog. Between May and August that first year, as I figured out how and where to find them, I found 26 new localities for the frog. Working from soil and topographic maps to identify potential hillside and stream bottom seepage wetlands, I located 56 new localities in the second season, including the first 3 localities in Alabama. The following season added another 35 localities, bringing to 117 the total localities added since federal listing. By then, it had become apparent that the Pine Barrens Treefrog was much more common and widely distributed than had been thought at the time of listing. At the request of USFWS, I prepared a delisting package for the Florida population of the Pine Barrens Treefrog in 1980, and the frog was ultimately removed from the Endangered Species List in 1983 (USFWS 1983).

Florida designated the Pine Barrens Treefrog as a Species of Special Concern, and monitoring of these newly discovered populations continued. On 21 July 1982, I was monitoring calling activity at localities on Eglin Air Force Base

(AFB). Protocol involved visiting sites after dark and listening for calling males. If none were heard, I would vocally imitate the call in hopes of soliciting a call response. At 9:18 p.m., I stopped at Milligan Creek and immediately heard several Pine Barrens Treefrogs calling south of the road. As I was writing in my notebook, I became aware of an unfamiliar call emanating from a small tributary stream next to the road. The call consisted of a series of 12 to 20 "chuck" notes repeated at about 5 notes per second. It didn't sound like a bird, but it was unlike any of the frogs known from Florida.

I laid my notebook aside and went to investigate. I soon found a small ranid frog sitting in shallow water along the edge of a rivulet. It had dorsolateral folds and looked like a small, tan Bronze Frog (*Rana clamitans*). I placed the frog in a plastic bag and continued on with the night's Pine Barrens Treefrog survey. I heard the same odd call at five more sites that evening, and investigation at each site revealed a small, tan frog that looked much like a Bronze Frog.

I got back to the motel around midnight and pulled out one of the evening's peculiar frogs. The first thing I noticed through the plastic was that the webbing of the rear foot was very reduced. Three phalanges of the longest toe extended beyond the webbing, whereas in the Bronze Frog only two phalanges extend beyond the web. The third phalanx was as long as the first two combined, so the free section of the toe was double that of the Bronze Frog. The other four toes each had two phalanges beyond the web, whereas the web of the Bronze Frog extends to the tips of those toes. This was clearly not a Bronze Frog and not a species previously known from Florida.

However, I was troubled by one thought. Eglin AFB is a military installation, and from 1955 to 1975 the United States had been engaged in the Vietnam War (which the Vietnamese call the American War). Had some airman returned from Southeast Asia with a few frogs as souvenirs? I sent some tissues to Tom Uzzell, who soon confirmed immunologically that this frog was a member of the *Rana catesbeiana* group of North American frogs. There was no longer any doubt that it was a new species.

Surveys for this novelty in 1983–1984 produced 10 additional localities, all in small seepage-fed streams draining to the lower Yellow River in Okaloosa and Santa Rosa Counties, Florida. Seven of the new localities were on Eglin AFB, and the other three were along three small adjacent streams on private timber lands immediately across the river from Eglin.

I now had a series of specimens and had identified the tadpole. It was time to prepare the description for publication. For the scientific name, I chose *Rana okaloosae*, after Okaloosa County, where I first found the frog. The name *okaloosae* is shared with a fish (*Etheostoma okaloosae*) and a crayfish (*Procambarus okaloosae*) also native to the region.

But what should the common name be? Early on, my colleagues and I had begun referring to the frog, somewhat tongue in cheek, as the bog frog. Dale Jackson, with the Florida Natural Areas Inventory, began tracking occurrences of the species and labeled it as the "seepage stream frog." When I reached the point in the manuscript to suggest a common name, I wrote Dale's rather cumbersome "seepage stream frog" as the common name.

This was in the days before personal computers and word processors. The usual procedure was to write out the text in long hand on a yellow legal pad and give it to a typist to prepare the typed manuscript. I completed the draft and gave it to our typist, Terri Crown, to type. Two hours later as I was sitting in my office, there came a tap-tap-tap at the door. The door opened, and Terri stuck in her head and said with great conviction, "It's a Bog Frog." And that is how the "seepage stream frog" became the Bog Frog.

Hydrologists might argue that the habitats occupied by Bog Frogs are actually fens rather than bogs because they are maintained by constant seepage of water from surrounding uplands. However, "fen" is a term unfamiliar to most nonhydrologists, who, if they know the term at all, associate it with Wordsworth's poem "London 1802," in which he suggested that England had become "a fen of stagnant waters" (an oxymoron by modern usage, since fens are, by definition, not stagnant).

I long assumed that Terri had insisted on the name Bog Frog because it rolled so pleasingly off the tongue. There was a second alternative that I had not considered. I recently had lunch with Terri. It had been almost 40 years since she had insisted on Bog Frog as the common name for the species. When I reminded her of the day, she smiled and clarified that she had anticipated needing to frequently type the name in various correspondence and reports. She noted that Bog Frog requires only 8 keystrokes, whereas 19 keystrokes would be required for seepage stream frog. Thus, it seems that the Bog Frog may actually derive its name from the resulting economy of clerical effort.

References

Christman, S.P. 1970. *Hyla andersonii* in Florida. *Quarterly Journal of the Florida Academy of Sciences* 33:80.

USFWS (US Fish and Wildlife Service). 1977. Final Endangered status and critical habitat for the Florida population of the Pine Barrens Treefrog. *Federal Register* 42:58754–58756.

USFWS. 1983. Final rule to remove the Florida population of the Pine Barrens Treefrog from the list of Endangered and Threatened wildlife and to rescind previously determined critical habitat. *Federal Register* 48:52740–52743.

About the Author

Paul Moler grew up outside Atlanta, Georgia. Following graduation from Emory University, he moved to Gainesville, Florida, where he received his MA in zoology from the University of Florida in 1970. He joined the Florida Game and Fresh Water Fish Commission in 1977 as the first herpetologist in the Wildlife Research Laboratory. Over the next 30 years, he worked with American Crocodiles, Eastern Indigo Snakes, Alligator Snapping Turtles, softshell turtles, Pine Barrens Treefrogs, Bog Frogs, and sirenid salamanders. Since his retirement in 2006, he has continued "working" as a volunteer at the Wildlife Research Lab. He also regularly participates in herpetofaunal surveys in Vietnam and southern Africa.

Part II
ADVENTURE AND EXPLORATION

9

MY FIRST SUMMIT CAMP

Maureen A. Donnelly

I haven't forgotten many firsts in my life, and I'll never forget my first *tepui* summit camp in Venezuela (*tepui* is a Pemón word for the table-top mountains in the region). My first tepui trip was not to a typical tepui. My first summit camp was located on a narrow ridge across from Pico Tamacuari—a massive granitic cone in the land of sandstone sky islands. Forest canopy and mountains stretched out as far as you could see in every direction. As the helicopter hovered, we tossed our gear to the ground and then jumped out while dutifully following instructions to keep our heads down as the helicopter lifted skyward for its return to the base camp. My first expedition to Venezuela in 1989 blew the doors off all my prior experiences of "living the tropical dream."

I decided to become a field biologist when I was 19 and contemplating my future along the Río Cuchuhaqui in Sonora, Mexico. I earned my doctorate with Jay Savage, an expert on Costa Rican herpetology, and spent 22 months in a remote Costa Rican field station conducting my dissertation research. I'd been a student on an eight-week-long tropical field course in Costa Rica and taught that course a decade later for the Organization for Tropical Studies (OTS). Despite my OTS immersion in tropical biology, every part of my first expedition to Venezuela was novel. I had no way to anticipate anything, and it was exhilarating. The Tapirapecó Expedition was my first trip south of the equator.

I knew about the *Mundo Perdido* (Lost World) in Venezuela because Roy McDiarmid was the outside member of my dissertation committee. Roy sampled several tepuis with Charles Brewer, a Venezuelan adventurer, dentist, and tepui explorer. Roy's talks detailing his tepui explorations were riveting. He was with the field crew that was stranded for several days on Cerro Neblina when the helicopter rotor broke. Roy's lived reality seemed far outside anything I would ever do as a tropical biologist. I counted myself lucky to have grown up learning field biology in Sonora, Mexico, and then I lived the tropical dream studying a species of poison frog in a Costa Rican lowland forest. I never saw myself being part of a scientific expedition, let alone an expedition exploring such an exotic place. I never knew my postdoc to study frog diets at the American Museum of Natural History (AMNH) in New York City would redefine the "tropical dream" I'd been living since 1979.

The Tapirapecó Expedition to "*La Ultima Frontera*" (the Last Frontier) was a year-long effort led by the Fundación para el Desarrollo de las Ciencias (FUDECI) to explore a poorly known mountain range in the southern extent of Amazonas territory in Venezuela. The Tapirapecó range, east of Cerro Neblina, includes the headwaters of the Orinoco River and is home to the Venezuelan Yanomami. The AMNH part of the expedition was led by Charles W. Myers, who was my postdoc supervisor. I remember the day Chuck called in December of 1987 and invited me to participate in the expedition. I was amazed that the opportunity was mine—I had just defended my doctorate. I clearly was in the right place at the right time, and I was able to take advantage of the "once-in-a-lifetime" opportunity. I immediately accepted Chuck's invitation to participate, and immediately all my flaming "poser syndrome" fears crept in: Would I be able to collect efficiently? Would I get along with everyone? Would I contribute positively to the experience?

My first trip to Venezuela was my first field trip with the late John W. Daly, who collaborated with Chuck on poison frog discoveries. John, the world's expert on frog alkaloids, was always searching for compounds in the next amphibian skin, and he became one of my favorite field companions following our first tepui trip. One normally does not associate alkaloid biochemistry with tropical field biology, but John studied wildlife management as an undergraduate in Oregon, and he was one of the best field biologists with whom I ever worked. He would go anywhere in search of animals, had an uncanny internal compass, and loved climbing uphill as much as he loved fishing. John and I spent hours collecting animals that Chuck prepared as specimens back in camp.

Twelve of us left snowy New York City for Caracas on a Sunday afternoon in early March. Our colleagues at Universidad Central welcomed five of us to campus Monday morning. Our midday meal was sprinkled with academic conversations about future opportunities to collaborate. After bidding our colleagues *adios*, we returned to the Anuanco Hilton apartments—a complex adjacent to the next-door Hilton Hotel. We were excited for our trip but confused by dark plumes of smoke smudging the city streets at dusk. Furious students set stacks of old tires ablaze on city thoroughfares. The sudden increase in bus fares economically harmed the poor, and poor university students were way past angry. The president of Venezuela declared martial law that night and instituted a dusk-to-dawn *toque de queda* (curfew). It was the first time I'd been placed under martial law; it was the first imposed curfew I'd experienced as an adult, and I recalled the fear I tasted while watching the movie *Missing* as an *extranjera* (foreigner) in Costa Rica—beyond grateful for no standing army in the "Switzerland of Central America."

I'd seen armed military men barely out of boyhood—in Mexico, in Guatemala, and in San Salvador, but I'd never directly experienced the use of military might while in a foreign country, and it was unnerving. The curfew meant we were sequestered in our apartment rooms in the Aunauco Hilton: watching citizens swarm stalled vehicles on the highway below our building, watching businesses being shuttered on TV news, and watching military personnel enforce calm upon the citizens of the nation. Tanks were parked in front of the Hilton Hotel. I was reminded of the Watts riots the day after I turned 11 and watched my hometown burn.

Our expeditionary team was headed to the southern end of the Federal Territory of Amazonas controlled by the military, and military personnel were busy holding down civil unrest for the president. Three of our colleagues, unnerved by the violence and militaristic response they observed, left under cover of night and flew back to New York, describing the scene as being a "bloody Beirut." Chuck spent considerable time on the phone with museum administrators defending his decision to go forward with the expedition. Nine of us waited in Caracas. From the privilege of our apartments, we watched the local news, cable TV movies, or the Westminster Dog Show as we sipped *cuba libres* (rum and cokes) after dinner. During the day when the curfew lifted, we swam laps in the pool at the Hilton Hotel, tried in vain to gather supplies at shuttered stores in heavily guarded malls, and ate our buffet meals in the Hilton Hotel bar while we waited. Days of waiting ensued. I understood waiting from doing so in several Latin American nations for one official reason or another—I knew that getting agitated was not going to speed the process.

I kept my eye on our expedition leader, and if he was cool, I could be cool too. I'd seen the glint of campfire licking the barrels of automatic weapons cradled by Mexican *federales* looking for cigarettes—I understood how to stay calm. My first experience in Venezuela's Lost World—so named by Sir Arthur Conan Doyle—was literally starting off with a bang.

Conan Doyle was inspired by lectures he heard in London as the earliest explorers returned home to recount their experiences on Mount Roraima—a tepui that sits where Venezuela, Guyana, and Brazil meet. Sir Arthur, famous for his Sherlock Holmes stories, populated his tepui with dinosaurs. His explorers climbed the mountain as did all early teams of explorers. Walking up the mountains meant there was little time for summit exploration—provisioning basic supplies was complicated and heavy. Helicopters are the great equalizers and allow modern-day explorers vastly expanded access to sheer-sided sky islands. While these expensive flying machines make the improbable possible, they can be dangerous, and I've met good pilots who later died in helicopter crashes. Every pilot I flew with flew by sight, and flying in a land of sheer mountains shrouded by clouds was often scary.

My insatiable curiosity means that new places and new people will immediately captivate me, and the jungles of Amazonas were no different from anywhere else I'd traveled. In the Mavaca Base Camp, I traded a pair of my shorts for a hand-crafted bow and three arrows—they hang on my living room wall in Miami. I walked along Yanomami foot paths in the forest near the base camp while searching for animals. I purchased Yanomami baskets—dyed red with achiote and accented with black—these mementos also hang on walls in my home. I was fascinated by the Yanomami—especially when one of the men tapped my cheek as he asked Raimundo, a FUDECI staff person, if I was from up river. Raimundo shook his head, said no, and asked why. The man replied, "Her eyes are the same color as theirs."

Every memory of my first expedition is an indelible one: my first meeting with the only woman I know who climbed tepuis (Kathy Phelps), my first ride in a military transport plane, my first exposure to the Yanomami in Caracas on the military transport plane, my first time in a helicopter, my first taste of bush meat, the first time we tried, and failed, to communicate by radio with the Mavaca Base Camp, the first time I watched a helicopter salvage bodies from a plane that crashed in the forest below our ridge, and the first time I saw the Southern Cross in the darkest sky I'd ever seen.

Every experience was unique—like making a snap decision to take a malarial drug prophylactically just after we arrived in Caracas even though it could make your skin turn black and fall off, being able to use my broken Spanish, learning bad words used in Venezuela, sleeping in hammocks, flying over rivers snaking through unbroken forest, viewing for the first time the unusual non-sandstone element of the Lost World, exploring the Last Frontier in Venezuela, seeing male Cock of the Rock, hearing a tapir crash through the understory, losing 10 lb as we ran out of food, being the base camp's cigarette lady, catching pirañas with the best ichthyologist in Venezuela, and spending hours and hours collecting amphibians and reptiles with John Daly.

Our expedition-issued radio did not work, except to receive AM radio stations from Brazil, so our group of six was truly alone on the ridge next to Pico Tamacuari. The other members of the expedition were on a different mountain or collecting fish in the base camp. The immense gray cone defined one side of a steep-hanging valley, and our ridge marked the other side. The beautiful stream running through the valley transformed into a raging torrent during heavy downpours. John, the ornithologists, and I were trapped on one side of the rushing waters the first afternoon we spent exploring the valley. John was the first to finally cross the stream, and the rest of us followed him after an hour of being soaked by the unrelenting rain I feared might never stop.

John and I roamed all the trails that FUDECI established on the Tamacuari ridge, and John cut new trails. We walked to the "tongue of the tapir" that marked the southernmost point in Amazonas Territory according to our maps. As we walked the trails by day, we could hear frogs calling from small caves and grottos all around our summit camp. On more than one occasion John held me by my ankles as I tried to catch these males that would retreat and cease calling the moment a headlamp's beam revealed them. We struggled to gather our meager sample of terrified frogs. One night after an afternoon rainstorm, I was plucking sleeping frogs off vegetation about 1 m above the ground. The next morning, much to our surprise, we discovered these common sleeping frogs were the same species as the "hell frogs" we found calling from caves. *Ceuthomantis cavernibardus* (Myers and Donnelly) was a species new to science named as a species of *Eleutherodactylus*.

As our expedition ended, most of our group decided to return by river, but I had to leave for my first job interview in central Florida. John needed to return to the National Institutes of Health, and the ichthyologist had to get home. As we were leaving Mavaca Base Camp with thoughts of real meals dancing in our heads, Chuck said to me, "I thought you might be good at this, but I never thought you'd be this good." My "poser" fear, that my first tepui camp would be my last, evaporated like the early morning mist over the Mavaca camp.

About the Author

Maureen A. Donnelly is a professor of biological sciences at Florida International University (Miami, Florida). She received her BA in biological sciences at California State University–Fullerton in August 1977 and her PhD from the University of Miami in December 1987. After teaching embryology at University of Miami (spring and summer) 1988, she moved to New York City to work at the American Museum of Natural History (postdoctoral fellow and curatorial assistant). She returned to Miami for a second postdoc in 1992 and was hired by Florida International University (1994). She served as the secretary (2000–2015) and president (2016) of the American Society of Ichthyologists and Herpetologists.

10

DOWN UNDER

Danté Fenolio

This quote, from H. P. Lovecraft, one of my all-time favorite authors, struck me: "The oldest and strongest emotion of mankind is fear. And the oldest and strongest kind of fear is fear of the unknown." The idea that the unknown presents real fear to the average human is something I've thought about when exploring the remote places I've worked around the globe. While some people fear the unknown, poorly explored places oftentimes have the most interesting biology. And I've realized something else: while remote places represent the unknown, they aren't to be feared; they are to be treasured.

As a boy I thought few things as exciting and mysterious as one particular type of remote place. My grade school teachers remarked on the exuberant enthusiasm the subject commanded of me—much to their dismay, something my schoolwork couldn't muster. The simple mention of one of these places gave me butterflies of the variety found in high school football players before a big game. Culturally, they are *incongruous*; Hollywood glamorizes them, yet protective parents deem them "off limits." They are a paradox in having some characters with quaint simplicity and others with amazing complexity. They exist in a dimension where time isn't measured in minutes, hours, days, or even years but rather by the development of stalactite and stalagmite gardens. They are places enveloped in perpetual darkness with maze-like corridors, water-sculpted surfaces, otherworldly inhabitants, and endless possibilities. These possibilities were enough for the imagination of a young boy to manufacture a childhood's fill of adventures exploring them. Etched at the top of my youthful list of "must explore" places, caves had it all and then some. The stars of the show for me, of course, were, and are, the mysterious, white, and blind inhabitants of the underworld.

So alien to mankind are the denizens of the subsurface that initial human reaction to them is extreme. Our imagination is challenged such that they might as well be entities from another planet. Glaring evidence of our amazement and bewilderment is reflected through the names we give them. As a group, groundwater dwelling organisms were initially designated as "stygobites" and obligate cave-inhabiting terrestrial organisms as "troglobites." The Greek-based name "stygobite" ties in with the River Styx, a mythical waterway of Hades, home to the dead. Some stygobitic amphipods (shrimp-like invertebrates) in the Ozarks are included in the genus *Stygobromus,* or "food from the river Styx." This is an appropriate name considering that amphipods probably constitute an important dietary component of cave fishes and cave salamanders. Locals of the Ozark Plateau were so taken aback in their encounters with white cave salamanders that they named them "ghost lizards." It hasn't simply been in North America that human reaction to subterranean life forms has been extreme. For example, centuries ago, eastern Europeans stumbled across stygobitic salamanders that had washed from the mouths of caves. Awestruck, they promptly classified the creatures as "larval dragons." Scientists later described this as Europe's famous groundwater salamander, the Olm (*Proteus anguinus*).

Underground inhabitants have not gone unnoticed by writers. Utilizing settings in the underground environment to add exotic and mysterious flare to a story, who could forget J. R. R. Tolkien's cave-inhabiting creature Gollum and the blind cave fish upon which he fed? Lovecraft wrote of troglobitic penguins in his *At the Mountains of Madness.* He described white birds with only vestiges of eyes living in dark caverns built by an ancient race. Michael Mott once wrote, "The legends, myths, and literature of mankind have always been filled with fanciful or terrifying accounts of underground lands and races, hidden from surface sight." All these authors conveyed the mystery and sense of adventure people experience and feel when exploring subterranean habitats.

My investigation of caves and their inhabitants has spanned the United States. I spent three years studying the population ecology of the Grotto Salamander (*Eurycea spelaea*) for my master's degree. This exposure to cave wildlife fueled the fire. Cave work quickly extended to far-flung places such as China, Thailand, Eastern Europe, Brazil, and Central America. Funding agencies for cave bioinventory work included organizations such as the National Geographic Society, US Fish and Wildlife Service, and several state wildlife agencies. Inasmuch, I strove to improve my photographic approaches while working underground to produce impressive reports and other deliverables. I started by developing my own very crude (but effective) photographic aquaria—systems that would allow me to focus on the detail of the species I was encountering. I also worked to improve my overall photographic skill set, learning more about flash photography and subtle lighting effects. Finally, I invested time in trying to find the right camera rig so that I could photograph aquatic wildlife in cave rivers, streams, and lakes, where and how I found them. This involved an underwater housing for my camera and large strobe flashes mounted on arms that extended from the housing. Several camera rigs later, I will say that spending time in subterranean waterways, photographing the wildlife that lives there, is easily one of the best things I have done in the field. I absolutely love capturing on film stygobionts and troglobionts where they live. Getting to a place at which I had a functional underwater camera rig involved a series of "proving trips."

One of my "proving trips" for a new, smaller aquatic photographic rig took place in a large subterranean river in Tennessee. I was assisting a good friend and colleague, Dr. Matthew Niemiller, with a population count of federally endangered Tennessee Cave Salamanders (*Gyrinophilus pelleucus*). The cave river meandered through a spacious cave with large pools connected by running streams. I slowly made my way into the largest of the cave pools and floated out toward the center, scanning the bottom of the pool for salamanders. This was the first time I had tried to photograph this species, in situ, and I wasn't sure how hard it would be to find them while floating at the surface and holding a camera rig. With this smaller and lighter rig, and as if on command, I floated toward a submerged rock ridge and a large adult salamander sitting at the edge of the ridge. I slowly aimed the camera and began shooting. The salamander did not react, and I was able to take a series of images from different angles. Advances in technology had made my life easier. The size of this rig, much smaller than previous versions, was much easier to maneuver.

Another "proving trip" involved advances in viewfinders and being able to focus a camera rig while looking at the back of a camera, through the aquatic housing, and from a distance. This proving trip took place in Slovenia, while trying to photograph the "human fish" (*Proteus anguinus*) in situ. I was fortunate to have a couple of friends from Slovakia who were able to plan a trip for me (to Slovenia) in order to photograph these outstanding amphibians in a large cave river. Most of the opportunities to shoot the salamanders would be in a meter or less of water. To reduce the disturbance and vibration of my movements, hunching over and simply holding the rig, rather than lying down in the water, worked out far better. Very slow and methodical movement of my feet and legs did not bother the salamanders, and I was able to get within shooting distance using the technique. The view finder on the back of the rig, through the aquatic housing, was a major step up in size and clarity. I was able to focus on these ancient groundwater amphibians while looking at the viewfinder held below the surface of the water.

Yet another set of "proving trips" had more to do with further development of my photographic aquaria than with use of an aquatic camera rig. These units became smaller and more portable, and I added features to help with the proper exposure of organisms that had little pigment. For example, the Georgia Blind

Salamander (*Eurycea wallacei*) is one of the spectacular stygobionts in Florida and Georgia. Most of the caves where this salamander is found are completely submerged, so cave diving is required to access the habitat. The research group with which I have been a part works with a team of cave divers, and they were ideal for our studies of this amphibian. I needed to document details photographically of the animals that the cave divers brought to the surface. I had been working for years on tweaking the design of a standard "v-tank." White and pink life forms can be difficult to photograph and are easy to overexpose. I needed aquaria where I could control the lighting and make adjustments. The final design for the aquaria I use today, and used with images of *E. wallacei* breeding in the lab, provided the flexibility I needed and allowed for the salamanders to sit, motionless, in a position while I made adjustments. One final benefit is that I got proficient enough with these aquaria that I could shoot relevant aspects of a specimen and have everything wrapped up in less than five minutes. Comfortable and flexible designs to these systems allowed for decreased handling time—something we all wanted to see. These proving trips and laboratory exercises were a huge success.

Perhaps the only other real proving trips came with designing a photographic rig that I could use belowground when we were exploring terrestrial cave habitats. I wanted something lightweight but powerful enough to deliver a high-resolution image, appropriate for publication. Advances in photographic technology have provided smaller, lighter rigs that produce images with a resolution that dwarfs older, larger rigs. The same can be said for flash technology. So these proving trips were simply a matter of testing various rigs to see how comfortable they were when crawling through caves or climbing rocks, walls, or ropes. The rigs allowed for the documentation of other herpetofauna in caves, sometimes species there as "incidentals" but sometimes not. For example, there was an afternoon in south China when we were surveying a cave for the first time. The waterway was the only way into the system. This water-worn cave passage had perfectly formed stone shelves along its walls that provided damp sites of refuge for cave life near the water and then dry hideouts higher above the water. The first dozens of these shelves had nothing living on them, but as we wound around a corner, something moved on one of these shelves. It was a coiled snake. The Asiatic Watersnake (*Trimerodytes aequifasciatus*) commonly inhabits cave waterways in Southeast Asia. These snakes are rear fanged and "mildly" venomous, focusing on fishes as prey items. Interestingly, we have found them in many caves with blind cave fishes and have wondered if they consume the obligate groundwater fishes. Probably so. The snakes seem to do well, even in deep reaches of caves, and I was able to use a smaller and lighter rig, perfect for photography while crawling through a cave, to capture images of the species in situ.

I've had the great good fortune to meet a host of spectacular cave-obligate amphibians in the process of testing my gear over many years of crawling through caves, swimming through subterranean rivers, and hunting through the rocks of cave streams and pools. As a byproduct of the efforts, I have amassed a nice image library of subterranean fauna. I also discovered an activity that I connect with at my core—snorkeling subterranean waterways. I hope that the image set intrigues the public enough to contribute to conservation efforts of these systems and the species that call them home. Groundwater contamination and overextraction, as

While it took multiple attempts to develop the right mix of lighting for a "half and half" image in a cave waterway, since I figured it out, I've used the technique many times. I like that the viewer gets a bigger picture of the environment, both above and below the water. These images convey the special scenes that I have been able to observe for decades; this is truly bringing the cave environment, and subterranean wildlife, to those who will never see it in person. My hope is that by seeing these special places with their highly adapted species, people might start to value subterranean ecosystems, ultimately acting to protect them. In this image, an Appalachian Cave Crayfish (*Orconectes {Cambarus} packardi*) was photographed crawling by a side passage in the cave that extends off into the distance. It was photographed in Kentucky, USA. Photo by Danté Fenolio.

well as other activities that lead to the wholesale removal of subterranean habitat, now threaten many subterranean species. They desperately need our help.

About the Author

Danté Fenolio was hired to develop a new conservation program in 2013 and is currently vice president, Center for Conservation and Research (CCR), at the San Antonio Zoo; director of research for the Amazon Conservatory for Tropical Studies (Loreto, Peru); and a visiting scholar at Trinity University (San Antonio, Texas). Active projects range from deep-sea work in the Gulf of Mexico to field projects in Japan, China, Chile, and Peru. He is the author of *Life in the Dark* and is a coauthor of *Cave Life of Oklahoma and Arkansas* and *Cave Life of Tennessee, Alabama, and Georgia*. Lab work at CCR has produced first-ever captive breedings of Georgia Blind Salamanders, Dougherty Plain Cave Crayfish, Comal Springs Riffle Beetles, and Reticulated Flatwoods Salamanders.

11

LESSONS FROM THE FIELD
It's the Journey, Not the Destination

Karen Lips

We need field biologists more than ever. We are losing species faster than ever to a range of threats, including habitat loss, invasive species, pathogens, and illegal harvesting. Yet one has to spend only a few minutes with the International Union for Conservation of Nature Red List to realize how little we know about the most basic facts for most species on the planet. There are many reasons for this—the overwhelming numbers of species to study, remote and difficult field conditions, and insufficient and dwindling research infrastructure (e.g., limited funding for exploratory research, different priorities in the academic reward system, loss of organismal biology courses in university curricula, reduced numbers of field courses, and funding pressures for field stations).

There are many reasons we should double down on supporting fundamental organismal biology: if we are to manage sensitive species properly, we need to know demographic rates, seasonal cycles, and habitat use; if we are to conserve them, we need to know historic trends, biotic interactions, and distributions. If we are to use them as the basis for ecotourism, inspiration, engineering design and insights, or other goods and services, we need to be able to find and observe them. Although technology is rapidly expanding our ability to observe remotely, field studies are still one of the best ways to gather these kinds of data.

Picking the right place can provide loads of new discoveries, interesting natural history observations, and a lifetime of stories. As a graduate student I was fortunate to have worked at a remote site in the Talamancan Mountains of southern Costa Rica. Later I set up camp at other sites across the mountains of western and central Panama and eventually made it to the mountains of eastern Panama on the Colombian border. I learned a lot from living in those remote sites, lessons that serve as a source of inspiration, laughter, and caution.

One trip in particular stands out for many reasons. It was 1990, and my advisor, Jay Savage, had asked if I wanted to conduct some biotic surveys at Finca Las Alturas, a site in southern Costa Rica where Stanford University would eventually build a field station. I was there to explore the region, develop a list of amphibians and reptiles at the site, and identify candidate sites for amphibian monitoring following the numerous recent reports of the mysterious disappearance of amphibians. In July I was joined at Las Alturas by Craig Guyer and Sharon Hermann, and we decided to push farther into the mountains and spend a few days collecting at a place called Valle de Silencio.

I had heard about this place from Luis Diego Gomez, director of the Organization for Tropical Studies' Las Cruces Botanical Gardens in San Vito, Costa Rica. Luis Diego was a botanist and explorer who had recently given a presentation on Costa Rican biodiversity and showcased the natural wonders of the Amistad Biosphere Reserve (one part of Parque Internacional la Amistad), a huge binational Peace Park on the Costa Rica–Panama border that protected the oak forests and the watersheds of the Talamancan Mountains. This park was as far from the capitals of San Jose and Panama City as you could get and had not been explored much from either side. Luis Diego was an incredible storyteller, and in his presentation he showcased Amistad as an important site of Costa Rican biodiversity and highlighted extensive tracts of old growth cloud forest, large numbers of interesting and endemic species, and untold numbers of species yet to be discovered. I asked him about his treks into Amistad, and he described a recent trip to Valle de Silencio, a high-elevation bog near the continental divide somewhere near Cerro Frantzius. He described a well-marked trail starting near a park station that covered an elevational transect through primary forest that was a long day hike; the trail sounded relatively easy if somewhat vague on location details. He was effusive in his descriptions of the amazing plants he found, the abundance of caecilians, and the numerous signs of tapir, details that were too much for us to resist. A unique habitat in an unexplored biodiversity hotspot with abundant caecilians? As Craig said, that was like finding gold for herpetologists. We were hooked.

Based on the fragmentary information from Luis Diego, we expected a long day hike, a day or two of collecting and exploring the bog, and then a day hike out. The area was extremely remote, so we were prepared to drink from springs and creeks and to survive on canned food, bread, and crackers, knowing that pizza awaited us at the end. We gathered our camping gear, bought food, and left early one Saturday morning for Agua Caliente, a small town at the end of the road where the ranger station was located. We parked a ways out of town, as we doubted our small Samurai jeep would make it through a giant mud pit where the road into town crossed a small river. Plus, we were worried about how little gas we had as we had been unable to fill the tank that morning. Once in town we used our rudimentary Spanish to ask for information on how to get to the park and the particular trail Luis Diego had described. Luckily, a kind man named Don Julio offered to guide us partway up the mountain and use his horses to carry our heavy backpacks. Hours later he left us and our packs at an abandoned farmhouse at the edge of the park where the trail began. We asked Don Julio how far it was to the bog, and he simply pointed up the trail and said "*Más allá*" (farther). We discussed a plan for him to return with the horses on Tuesday (*jueves*) to guide us. This part of the trail had been more open, hotter, and more rugged than we expected, and it had taken much longer to get this far, so we stopped for the day.

The next day we headed out, hiking for hours along an old trail through a beautiful old growth oak forest, periodically checking under logs and moss for salamanders, lizards, and frogs. We were finding dozens of salamanders with little effort, but we saw no sign of a bog, or caecilians, and we had not crossed any streams, so none of the endemic stream frogs either. We realized that we must be walking along a ridge far above the river, which was not promising for either stream frogs or drinking water but explained the name of the valley. We stopped for the day and could hear a river far down the hillside. Craig set off to get water. When Craig returned he mentioned how far he had to go for water, and we began the mental calculus of all trips—how much food and water did we have and how far did we have to go? We did not have a map, so we did not know exactly where we were. Given our elevation we did not think we were close to Valle de Silencio, and we had only a small amount of food left. It was neither the first nor the last time I would realize that things often take longer than you expect. We were also concerned about getting back in time to meet up with Don Julio at the farmhouse on Tuesday and reluctantly made the decision to turn back the next day. That night we experienced a drenching rainstorm typical of the cloud forest that required emergency digging of a ditch around our old and leaky tent to keep us dry. When we arrived at the farmhouse on Tuesday, Don Julio and his horses were nowhere to be seen. As I replayed our Saturday conversation with Don Julio, I realized we had said *jueves* (Thursday), but we had meant *martes* (Tuesday). Don Julio would not be there for another two days.

After a quick rest we lifted our packs and hiked back to Agua Caliente across the many streams and deforested hillsides between the park and town. It took us much longer than we expected, but we arrived at the ranger station that afternoon hot, hungry, and tired. We found Don Julio and explained we wouldn't need his help on *jueves* but arranged to have him haul our packs out to our jeep on his horses. Before we left, we asked around town if anybody might have some extra gas and were able to purchase a few gallons of gas from a guy with a rusty 50-gallon drum in his backyard. We were exhausted, and hiking through the deforested areas in the blazing tropical sun made the hike to the jeep pretty miserable. Luckily, we were able to get the jeep started and drove it across the small river and back into town to get the gas, but now we were worried about the rainclouds spilling down from the mountains. The guy quickly siphoned a few gallons of rusty gas using his mouth and a rubber hose out of the 50-gallon drum as the rain drops started; we drove out of town in sheets of torrential rain. The jeep made it halfway across the mud pit before it got stuck. We knew we were running out of time, with every minute of rain converting the sticky mud into an inescapable trap, reducing our chances of getting home that night. We got out and pushed the jeep as hard as we could, getting sprayed by clumps of mud as the little jeep spun its wheels. Finally it gained purchase, and we drove up onto firm ground. We drove to town, where we sat in our stinky, mud-splattered clothes and ate the best pizzas we'd ever had.

Today anybody can take a guided ecotour along that trail, crossing the continental divide and lodging in *refugios,* and several collecting expeditions have been staged by scientists from the Universidad de Costa Rica throughout the Talamancan Mountains. None of us ever made it to Valle de Silencio, but I've often thought about that crazy trip. As it came during my very first field season, this trip made a lasting impression on me, and I learned many important and enduring lessons.

Communication: I think of this trip every time I plan any event in Spanish and do an immediate check on both sides of the conversation. I double check all my requests for information and directions, rephrasing my questions in different ways to see if I get the same answer. If possible, I ask multiple people the same questions and triangulate on the true answer. Regardless of how good my planning might be, it is always best to assume that it will take much longer than I expect, and I always double time and distance estimates and adjust my food, water, gas, and other logistics accordingly. If it sounds too good to be true, it probably is. At the time neither Craig nor I questioned Luis Diego's description, but we have laughed for years about mystical places where caecilians are dripping from the trees and tapirs are prancing in the forest.

The path less traveled: If you want to see new things and experience new adventures, get off the beaten path. When you do, you might not know exactly where you are going, but be safe and able to retrace your path. I can think of many times when I would rent a four-wheel-drive truck and drive up the dirt roads into the mountains as far as I could safely go, searching for new field sites and access points. What I learned was that the sites with good roads and easy access generally led to deforested slopes and human habitations. With bad, muddy, or difficult roads you risked getting stuck or lost and they could only get you so far, but they almost always offered promising glimpses of good habitat at the end of the road.

Safety first: Safety starts with having situational awareness of potentially dangerous people and dangerous conditions. Working in the rainforest will always be risky because of the variety of biological and environmental dangers, such as treefalls, predatory animals, venomous reptiles, biting insects, and dangerous microbes. Be sure you have a full tank of gas and a full water bottle before you leave, pay attention to where you walk, and carry flagging tape to mark new trails. I've walked by many large pit vipers that I never saw until the last minute, and I've been stalked many times by big cats. Luckily the snakes never moved, and none of the cats ever approached, even when I was living in a pasture where a jaguar was killing the young calves. I was terrified at the time, but I believe that, in general, greater danger is found in two-legged beasts than those with four or none. All my close encounters with dangerous people have been in cities, or beach towns where tourists are targets for robberies, not in the rainforest. The strategy for a woman traveling alone in remote places is not the same as it would be for Indiana Jones; for us, survival depends on fading into the background, not making eye contact, and not being noticed.

It's the journey, not the destination: My notes from the summer of 1990 are missing, and I have had to recreate what happened from my unreliable memory, comparing notes with Craig, and the official notes in the collection record. I don't remember when or where we stopped each day, why we did not push forward, or why we didn't have enough food. What I do remember is how I felt: the fun we had, the exhilarating sense of adventure, the hilarious realization of how we screwed up, the teamwork to extract ourselves, and how awesome that pizza tasted. As a field ecologist, I have lived at many sites for months or years at a time or returned annually over many years. I have seen how different species changed activity and abundance over time and the immutability and the adaptability of nature. I also saw how as the areas were developed, the roads improved, human populations expanded, and access increased. I came to know the people who lived and worked nearby, and I understood somewhat better their lives in these small towns and appreciated our similarities and differences. Although I never made it to our destination on that trip, my journey as a field ecologist has been incredible!

Dedicated to Jay Savage, Luis Diego Gomez, and all the other tropical explorers who came before and to all the explorers who will come after.

About the Author

Karen Lips is professor of biology at the University of Maryland. She has a BS in zoology from the University of South Florida and a PhD in biology from the University of Miami. She is an ecologist who studies how global change affects biodiversity of amphibians and reptiles in Latin America and the United States and the impacts of those losses on natural ecosystems. Dr. Lips was a Jefferson Science Fellow at the Department of State, where she worked in the Bureau of Western Hemisphere Affairs and served as an Embassy Science Fellow in Colombia. Dr. Lips has been a research associate at the US Museum of Natural History and at the Smithsonian Tropical Research Institute and is an elected fellow of the American Association for the Advancement of Science and the Ecological Society of America.

12

FLYING SOUTHWARD THIRTY-THREE DEGREES TO CATCH MORE FROGS

Ronald Altig

In 1975, as my seventh year at Mississippi State University approached, I planned a sabbatical in the Amazonian jungle near the equator at Santa Cecilia, Ecuador. Marty Crump gave me a brief introduction to this new world, and a couple of nights after she left I noticed the absolute brilliance of the stars reflected in a puddle of water on the dirt runway. When I lay down for a longer view, I saw a shiny satellite zipping across the backdrop of stars, and when my focusing knob got twisted just right, the Southern Cross snapped into view about halfway to the horizon. With the calls of several frogs providing background music, I realized just where I was positioned on my tiny home planet.

Ildefonso Muñoz's place, dubbed Muñozlandia, on the north bank of the Río Aquarico, had served as a research base for several researchers, and many frogs and a couple of plethodontid salamanders and caecilians were waiting to be found. Señora Blanca Muñoz and a Quechua girl named Ilysia fed me from a very interesting menu.

Humorous happenings arose during the resuscitation of my undergraduate Spanish. Recall that an *i* in Spanish is pronounced like a long *e* in English, so when a Spanish speaker chooses to use the common curse word that is an anagram of "this," it sounds like he is talking about a piece of paper, a sheet. One evening Ilde asked me if I needed some clean *sábanas*. I told him that I did not know that last word.

"Oh," he said, and explained with widely spread, upraised arms, "*Blanco y grande, ropas para su cama*" (white and large, clothes for your bed).

"Oh, okay," I responded, "sheets."

He frowned his disgust and admonished me loudly with "No, Ronn! Not is shits!"

"Si!" When I spelled it out loud, he mused more quietly.

"Ahhh, sheets, a good to know!"

Because nothing was known about the life history of *Nyctimantis rugiceps*, a large, garish hylid that breeds in bamboo stumps, it was a prime target. Their calls told me that they were not rare, and one night I quietly worked through a bamboo thicket until I could feel the vibration of a calling male when I lightly touched a specific stalk. I marked the stalk so that I could return the next day with an axe. When I got back on the path, I turned around to see where I had been, and right behind where I had just stood there was a *Nyctimantis* sitting on the edge of a bamboo stump. I made the very painful decision to leave it alone with the hopes that it would lay eggs, and when I checked the next morning, there was a large number of eggs in the water in that stump—all dead! I never found another adult.

About a 10-minute walk west through the jungle, there was a small forest pond dubbed Lago Perdido (Lost Lake). It was about 40 by 20 m, and many frogs bred there. I avoided the western end because there was a small grove of palms with hordes of long, narrow spines, and several young *Caiman crocodilus* sunned along the edges. I caught a male microhylid *Ctenophryne geayi* strictly by accident. The bank of the pond had a very thick deposit of leaves, and one evening I was sitting there waiting for it to get dark. I jumped when a frog called from beneath the leaves right between my feet and my butt. I heard it call several times until I saw a leaf vibrate about four layers down before I grabbed it.

Phyllomedusid frogs can grasp with their hands and feet, and the frantic ovipositional antics of the skinny legged *Phyllomedusa vaillantii* on leaves above the water were quite comical. Their oddly shaped tadpoles with a mobile tail tip and a ventral fin taller than the dorsal fin hung in midwater in a nose-up posture. On one night, a weird "microhylid thing" happened that I eventually saw repeated only one other time with a species of *Microhyla* at a similar pond in Vietnam. Many individuals of one species of *Chiasmocleis* were moving into the pond from one direction, and before I left that night a large area at the western end of the pond was covered with their floating films of eggs. I still have no idea why either of these migrations occurred on one specific night.

I sometimes walked out on a fallen tree trunk that spanned the width of the pond. *Phyllomedusa* tadpoles were hanging in the water, and strange, eel-shaped beasts stood upright with their tails stuck in the bottom. I bought some small fish hooks the next time I was in Quito, and one of those beasts instantly took the bait. I eventually found out that it was a synbranchid eel; the gill slits that are fused into one midventral opening was the diagnostic trait. Just as I have often wondered about *Siren* (two-legged aquatic salamanders) in small, isolated ponds in Mississippi, I wondered how these fish got into that isolated pond.

One night on my way back from Lago Perdido, I walked out into a long dug-out canoe that was beached on the mucky bank of a woodland lake. As I looked along the canoe, I saw a turtle pushing aside the plants on its way to the surface to breathe. I got balanced and ready, and luckily at the last moment, I realized that I was about to grab the head of a large electric eel (*Electrophorus* sp.)! It made a loud gasp when it took in a breath at the surface, and I expelled mine in return. Another time I walked down the sandy stream that flowed into that lake to check out some frog calls. The frogs stopped calling, so I shut off my light and waited. When I eventually turned on my light, to the side, there was a pair of large reflective eyes on the sandy bottom not far from my feet. I automatically assumed that it was a young caiman, but I finally realized it was a freshwater stingray. The manager of the parts depot for one of the drilling companies, whom I had met, had shown me the large toxic spine on the tail of one of these fish. When I jabbed it on the nose with my machete, it disappeared in a cloud of sand.

I never swam in that lake because I had been told that the piranhas would attack if they were breeding. I also never swam in the river because it was way too cold, and there were small catfish (i.e., candiru) that are parasitic on the gills of other fishes. They find their way into the gill cavity by being highly sensitive to ammonia, and small larvae can swim up a man's urethra for the same reason. One of the military officers suffered that fate while I was there.

Comments of admiration are added about some large, musical birds that often sang me awake at daybreak. A tall, remnant canopy tree behind my house was the nesting site of a group of Oropendolas. These large relatives of North American orioles warbled a constant choral of burbs, gargles, and farts while hanging upside down to weave their long, hanging nests.

Once in a while I would catch a bus toward Quito in the afternoon and then get off just before the Andean summit. The shiny, black adults of the bufonid *Atelopus ignescens* in a nearby stream looked like they had been carved out of licorice. Their strange tadpoles with bellies modified as suction cups were stuck all over the rocks in fast water. On my first visit to a smaller, steep stream, I saw nothing of interest except for a row of red dots aligned along the bottom of a small tree branch farther uphill. The next time I had stable footing, I was face-to-face with a large treefrog that was bronzy brown top and bottom, including its eyes. All its toe pads were bright red. Bill Duellman and I eventually described this frog in 1973, and it is now known as *Hyloscirtus lindae* in reference to Duellman's wife, Linda Trueb. When I got back on the road, a fellow from Chile with his strange Castilian pronunciations of Spanish picked me up. He gave me one donut of a dozen that he had for a girlfriend.

In 1973, John Lynch and Bill Duellman described several centrolenid frogs from a stream farther down the slope. Such large diversity in a single stream is uncommon, and one of those frogs was atypically brown. I saw several species of hummingbirds in their nightly hibernations on limbs over the water, but I did not see or hear even one centrolenid. I finally sat down and dosed against a tree until daylight. I had learned how to find centrolenid tadpoles in Mexico, so I walked along the lower, fairly level part of the stream until I found a leaf mat behind a small log that was projecting into the stream. Centrolenid tadpoles are elongate burrowers with tiny eyes, and many are bright red. When I dipped into that leaf mat, I saw many squirming red tadpoles that I could divide into five morphological types, but actual identifications were never possible. I did collect specimens of another hylid frog that looked odd to me. *Hyloscirtus torrenticola* is the other species that Bill Duellman and I described in 1973.

Bothrops atrox was reported to be the most common snake at Santa Cecilia, and I never saw even one. Frog people don't see snakes, and snake people don't see frogs! One day I heard a big commotion when some military officers were swimming in the river, and when I went to investigate, I found a young *Eunectes murinus* (anaconda) about 2 m in length. I was quite a hero when I grabbed it. Then I wondered what I was going to do with a large, heavy snake coiled into a ball around my left hand. I managed to uncoil it enough to hold it against the wall of my house until it relaxed with its tail touching the ground. I marked the position of its snout and thus had an estimate of its length. I took it back to the river and released it when no one was around.

The frogs at Santa Cecilia could be partitioned into those limited to primary forests versus those that occurred in disturbed areas. The hylid frogs in the *Dendropsophus leucophyllatus* and the *Scinax ruber* groups were common in disturbed areas. The long-legged hylid *Boana lanciformis* bred in the shower reservoir almost every night, so tadpoles in the soap were a herpetological addition to every shower. The microhylids and some phyllomedusids were limited to primary forests, and in both cases the frogs were found in specific areas that looked just like the rest of the habitat to me. The small green centrolenid *Teratohyla midas* was limited to a small grove of six banana trees growing in water along the runway, and I still have no idea why my attempts to rear that strange tadpole always failed. Likewise, the grumpy leptodactylid *Ceratophrys cornuta* bred in one small pool near the end of the runway. I concluded that the frogs recognized a lot more structure of tropical forests than I did.

I felt like I was decently adept at finding frogs, but there were some species that I never encountered. Who knows how populations of tropical frogs vary through time or how my searching methods biased what I found. I often book mental trips back to Santa Cecilia, but knowing that the area where I worked is now covered by an oil palm plantation definitely sours those visits. I cannot return to the past, but I hope the frogs have it figured out.

About the Author

Ronald Altig was raised outside a small farming community in central Illinois, and his education after high school included the University of Illinois (BS), Southern Illinois University (MS), and Oregon State University (PhD). He spent his academic career at Mississippi State University and retired as a professor after 30 years. He mentored 12 MS and 9 PhD students and published three books and 159 journal articles. Many trips around the world gave him many memories and a tantalizing appreciation of the biological diversity of many habitats.

13

TRIP TO THE XINGU RIVER IN THE AMAZON FOREST OF BRAZIL

Célio F. B. Haddad and Marcelo Gordo

In 1986, we were invited by Brazilian herpetologist Dr. Paulo E. Vanzolini to undertake a 30-day survey in the Brazilian Amazon Basin. Our mission took place in December of that year. The goal of our expedition was to survey the herpetofauna, with an emphasis on amphibians. This faunal information was required as part of the environmental report that precedes any construction work that significantly affects the environment in Brazil. We camped in the forest near the left bank of the Xingu River, near the municipality of Altamira, in the north of the state of Pará. We were three people on this trip: Dr. Adão Cardoso, professor at the State University of Campinas, São Paulo; Dr. Marcelo Gordo, then an undergraduate student and today professor at the Federal University of Amazonas; and Célio Haddad, master's student supervised by Dr. Cardoso and today professor at São Paulo State University. Adão was 35 years old and had extensive Amazon experience in the Brazilian state of Acre. Célio was 27 and had little Amazon experience. Marcelo was 19 with no experience in the Amazon rainforest.

What follows are some of our stories from that trip, meant to paint an honest picture of fieldwork in a remote area of the Amazon Basin in the 1980s. We found amazing diversity and abundance of amphibians and reptiles, experienced difficult and sometimes uncomfortable living conditions, and had some unusual cultural interactions—all memories we treasure today.

The company Consórcio Nacional de Engenheiros Consultores Sociedade Anônima, hired to manage the initial part of the prospecting work in the area where a hydroelectric plant would be built, had several camps. Prospecting refers to any service aimed at surveying the local flora and fauna as well as aspects of geomorphology. The camps were in clearings in the forest created for this purpose. They were covered by plastic canvas under which hammocks were installed in a common space shared by all. In these camps lived the workers hired by the company and the researchers who carried out the various surveys. About 200 people lived in our camp. In addition to our trio of amphibian experts, there was a group from the Institute of Technological Research of São Paulo doing analyses on the geomorphology of the Xingu River bed; their data were necessary for construction of the dam. Because the camp was covered in black plastic, it was unbearably hot in the middle of the day. To reduce the temperature, palm leaves were placed above the plastic, but even so the heat was intense. During the hottest times, we usually rested on a trail in the forest or stayed immersed in the Xingu River until the heat subsided a little. Insects added to the discomfort, especially in the camps, with mosquitoes at night and many horseflies and *piuns* (Simuliidae, black flies) during the day.

Camp workers were humble people but quarrelsome and rough. They walked around camp and in the jungle always with machetes around their waists. The machetes were used to open trails in the forest and for self-defense against wild animals, like jaguars—and also against other camp workers. We were always concerned and a little frightened when near them. One night, a worker called one of us (Célio) to see an animal that was near their hammocks. Célio (fearfully) obliged, and saw that the animal was the Giant Toad (*Rhinella marina*). Instinctively Célio grabbed the large toad and, to his surprise, noticed that the workers formed a distant circle around him and were wide-eyed and surprised at his courage in holding the "beast." Célio tried to approach them with the toad, and they ran away in terror. From that day on, the workers looked at us with admiration and respect, and we felt more at ease when near them. We even thought about always carrying a toad around our waists, to defend ourselves. The workers were afraid of toads because the animals are used in witchcraft in Brazil and because people believe that the toads cause diseases, like *cobreiro* (diseases that spread through the skin, such as mycoses, herpes, and others).

Food was sometimes a problem. On one of the first nights at the camp, we were served a barbecue dinner. We noticed a somewhat strange flavor in the meat, something that resembled a kerosene flavor, but it was mild. The next day everyone in the camp was sick with intestinal disorders that lasted throughout the day. We later learned that the meat had been contaminated with fuel from the helicopter that transported the food to the camp. When Cunha, the cook at one of the camps, finished cooking a 60-L pot of rice, he poured 2 L of soy oil over the rice and mixed it with a large spoon, claiming that this made the dish fluffy and delicious. In fact, the rice was quite greasy and cloying, but for the energy-intensive camp laborers, it was certainly adequate. Both the crackers, which came from Belém, and the cheese, made in Altamira, contained sand, which was really unpleasant to chew. Apparently they had been prepared with impure salt, which contained a lot of sand.

The workers with whom we shared the camp captured Yellow-Footed Tortoises (*Chelonoidis denticulatus*) in the forest and either restrained them in small enclosures or tied them to tree trunks by wires attached to holes they made in the animals' plastrons or carapaces. These tortoises were basically a live stock of meat that could be consumed whenever the men wanted. There was plenty of food in the camp, however, which did not justify the capture of these animals. We felt sorry for the tortoises, mainly because of the cruel way in which they were slaughtered for consumption, so we surreptitiously went to the places where the workers trapped the tortoises and released them all. The feeling of releasing the animals was both satisfying and thrilling, a good deed mixed with the danger of being discovered by the fearsome workers. Luckily we were never surprised by the tortoise owners, as this could have caused us a big problem!

Early on, we realized we needed a table so we could fix specimens of amphibians and reptiles and take notes. We requested that the person in charge of the camp, Manoel Correia, get us a large board to use as a table. He kindly offered to solve the problem. That same day we heard a loud bang in the forest near camp. We ran to see what was happening. Mr. Correia had cut down a Brazil nut tree (*Bertholletia excelsa*) over 30 m in height with a chainsaw to remove a board for our table. We felt a mixture of bitterness and anger for such a meaningless death of a beautiful, useful, and imposing tree, but we soon consoled ourselves that the tree and the entire forest were doomed by the flood that in a few years would be caused by the construction of the hydroelectric plant. This same logic consoled us when it was time to kill the animals we collected, as they would not have a promising future in this place condemned by the flood. Today, this forest has disappeared under the lake of the dam that stores the water for the operation of the Belo Monte plant, the fourth largest hydroelectric plant in the world. Certainly the remains of this Brazil nut tree felled to serve as our table are still under water, as it is hardwood, highly resistant to time and weather.

We also faced some sleeping problems. When it rained, it was common for leaks to wet our hammocks. The inexperience of sleeping in hammocks also affected us. We considered sleeping directly on the ground, but we gave up on that idea because of the large number of pit vipers, spiders, scorpions, and army ants that roamed through camp. The noise made by the workers also bothered us. They often played cards until dawn, shouting and fighting with each other, as those who lost part of their salary in gambling complained about the cheating of the winners. Sometimes we slept badly for three nights and on the fourth night we slept well owing to exhaustion.

One night in camp, a worker called on us to capture a snake. It was a pit viper, the Jararaca (*Bothrops atrox*), and was next to the camp's drinking water filter. One of the workers had almost stepped on the snake. The truth is that Jararacas were very common in that area, and we had several encounters with this species. One night we had traveled for a few kilometers on a dirt road in a Toyota four-wheel-drive truck. Célio was collecting frogs of the genus *Adenomera*, and fractions of a second before he grabbed a frog in a thicket of vegetation, the head of a Jararaca, also interested in the frog, appeared next to the frog. Célio almost got bitten by the snake. That night we collected several Jararacas. Because we didn't have wooden boxes to accommodate the pit vipers, we packed them in plastic bags. On the way back to camp, at dawn, when we drove on a small improvised bridge made of loose wooden planks, Adão lost control of the car and we overturned. Chaos ensued. In addition to the stress of the accident, the feeling of being under plastic bags containing live venomous snakes was really terrifying. Fortunately no one was hurt, but we couldn't turn off the engine of the overturned vehicle, so Marcelo ran for about 8 km to camp to ask Mr. Correia for help. He and some other men came to the accident site and rescued us.

The second part of the expedition took place in a much quieter camp, accessible only by boat, up the Xingu River, toward the mouth of the Iriri River, land of the Arara Indians. From this new camp we explored the area on a few nearby trails, but most of the time local assistants took us by boat with an outboard motor to different places. During part of our trip, Adão had commitments in Altamira and Célio had to return to São Paulo, leaving only Marcelo, the cook, and the cook's assistant in the new camp.

One morning in this new camp, Marcelo was tending to his sores caused by insect bites when the cook's assistant asked if he could bring a farmer, a neighboring resident, so that Marcelo could bandage this person's wound. Marcelo promptly agreed but noticed that the first aid kit was very simple. When approaching the injured man, who was drunk, Marcelo could see that he had two deep cuts in his leg just above the knee, made by a chainsaw. The wounded man was placed on a table, without his pants, where Marcelo could see the severity of one of the cuts, which was bleeding profusely. Luckily Marcelo had brought some thread and suture needles, but he had no pliers to facilitate handling this material. Without anesthesia and with only alcohol for cleaning, Marcelo started suturing the deepest cut, without gloves, in an attempt to stop the bleeding. Marcelo had difficulty getting the needle through the man's thick skin, causing him a lot of pain. The tension was so great that a puddle of sweat formed below

Marcelo, causing concern on the part of the "patient." The procedure took more than an hour, allowing the patient to tell his story as a fugitive from the Brazilian state of Maranhão, having killed a person in a fight. He explained that as a fugitive, now in the state of Pará, he worked as a professional hunter hired to supply the village of settlers nearby with food. He also told of having suffered an attack by the Arara Indians, his hunting companion killed at his side by arrows, and how he managed to escape by running through the jungle and throwing himself into the river to swim across; he was helped by the settlers on the other side. At the end of the suture work, there were 14 well-spaced stitches that successfully stopped the bleeding, and Marcelo closed the wounds with clean bandages and bactericidal ointment. The next morning, Marcelo arranged for another settler to take the injured man to the hospital in Altamira to be evaluated and treated with antibiotics and an anti-tetanus injection.

By the end of the trip, all of us were parasitized by chigoe fleas (*Tunga penetrans*), mainly on the feet but also on the hands. The sensation was a little uncomfortable, because in addition to scratching, it hurt. We removed chigoe fleas by making small incisions in the skin and squeezing to eject them. Black flies ended up causing many injuries to Marcelo's skin, resulting in a serious secondary infection and requiring hospital care as soon as he arrived in Campinas, São Paulo.

In addition to the beauty of the Amazon rainforest, the Xingu River, a tributary on the right bank of the Amazon River, is fascinating. With a length of about 1800 km, it is a mighty river with an average water flow of about 10,000 m^3/s. In the region where we worked there were the Xingu rapids which, in addition to the scenic beauty, contain rock boulders carved with enigmatic geometric drawings made by an ancient people that inhabited the region. Today, some of these carved drawings are submerged by the dam. Problems associated with modern civilization arrived at the Xingu River not only because of the flooding of the dam but also because of contamination of its waters by mercury, a result of the exploitation of gold by illegal miners that continues today in that region.

We collected more than 50 species of anurans. Some were new to science and later described by Janalee P. Caldwell, such as the Para Toad (*Rhinella castaneotica*) and the Brazil-Nut Poison Frog (*Adelphobates castaneoticus*). The great richness of species caught our attention, and the enormous abundance of amphibians and reptiles surprised us. The area was one of the most spectacular and remarkable places we have ever worked.

We dedicate this essay to the memory of Dr. Adão J. Cardoso, who led our adventure through the Amazon and died in a car accident in 1997.

The young student Marcelo Gordo handling an *Iguana iguana* captured on the banks of the Xingu River, Pará State, Brazil. Photo taken by the late Dr. Adão J. Cardoso.

About the Authors

Célio F. B. Haddad is a biologist and received his PhD in ecology from the State University of Campinas, Brazil. As professor of vertebrate zoology at São Paulo State University, Rio Claro, Brazil, he conducts research on amphibians in the areas of natural history, taxonomy, systematics, evolution, ecology, and conservation. Most of his fieldwork is done in Brazil, in tropical rainforest environments in the Atlantic Forest. He was research associate at the Museum of Vertebrate Zoology, University of California–Berkeley (1997) and visiting professor at Cornell University (2013). He is an Honorary Foreign Member in Herpetology of the American Society of Ichthyologists and Herpetologists and a member of the Brazilian Academy of Sciences.

Marcelo Gordo has a degree in biological sciences from the State University of Campinas, a master's degree in ecology from the National Institute for Research in the Amazon, and a PhD in zoology from the Museu Paraense Emílio Goeldi. From 1992 he has been a professor at the Federal University of Amazonas, teaching in the areas of ecology and conservation. He carries out research and conservation actions with amphibians, reptiles, mammals, and forest restoration.

14

WOK BILONG OL PIK

David Bickford

Exploration and discovery have always been fundamental to my fascination with the natural world. I relish being outdoors just looking for cool stuff. I have been extraordinarily lucky in how much of my time I have been able to do just that. While pursuing my doctorate, I was privileged to live and work on the world's largest tropical island, New Guinea, for nearly four years. That time, the people and places I came to know, and the experiences I had still inspire a deep sense of awe in me to this day.

Situated just north of Australia at one end of arguably the most interesting and important biogeographic transition on the planet (in the demesne called "Wallacea" after Alfred Russel Wallace, the cofounder of the theory of evolution via natural selection), the island is a field biologist's dream. After months of preparation in Coral Gables, Florida, at the University of Miami, I was set to journey to the Wara Sera Research Station inside the Crater Mountain Wildlife Management Area, situated between the provinces of Chimbu, Gulf, and Eastern Highlands in Papua New Guinea (PNG). I had built up an overload of excitement and anticipation at being able to conduct field research for my doctorate in one of the last remaining truly remote rain forests on the planet. My initial delight and joy, however, was soon to change.

The jungle in many parts of New Guinea is dripping wet: a haven for biodiversity of all kinds. With my field site receiving almost 7 m in rainfall annually, amazing adaptations possible only in hyper-humid conditions abound (e.g., the evolution of parental care in frogs with direct development). Terrestrial leeches do extraordinarily well there and seem to be omnipresent in Melanesian rainforests. We entertained at least seven different species of leeches on a daily basis, with unnerving abilities to enter orifices, latch on, and get a full feed without the host being any wiser. It was a veritable hothouse of other parasites and tropical diseases as well. Everyone fixates on the ones you can die from (e.g., malaria, lymphatic filariasis, systemic septic infections, and meningitis), but it was the leeches that had the largest impact on my daily routine. Minor nuisance though they are, since they do not transmit human diseases, they make you bleed profusely for extended periods because of potent anticoagulants. They also create itchy, bloody sores that easily become infected when scratched, leading to double whammy when coupled with the prevalence of a tenacious skin fungus ("grille," or *Tinea imbricata* cf.) that left open, festering fissures from toes to ankles. At one low point, I had dozens of open sores below my knees and a systemic septic infection that sent bluish streaks up my legs. The preponderance of novel bacteria, fungi, viruses, and parasites had my immune system on the run. My romantic idea of "fieldwork" had been replaced by a kind of dread and worry about my mental and physical health in those first few months of fieldwork. Luckily, and with a bit of help from my friends and assistants, I emerged after my "trial by infection" and had a wonderful, albeit challenging, experience.

One of my first memorable New Guinea experiences in the rain forest happened while I was recording frog calls near the field station. It was raining (surprise!) and I had my expensive recording equipment out to record the frogs. Back in 1995, we still used cassette tapes to record frog advertisement calls—a complicated affair compared with today's smaller, lighter-weight digital media devices and instruments. I had an umbrella in one hand, the recording equipment slung around my neck, a fishing vest and fanny-pack full of instruments, and a sensitive microphone in my other hand outstretched over a microhylid frog perched just a meter off the forest floor. These frogs don't need water for breeding and instead lay their eggs in shallow, cup-like cavities in the soil. This particular male was almost directly above his prepared oviposition site calling for all he was worth. I had to switch off my headtorch in order to elicit the frog to call and record without disturbing the frog too much (i.e., enough to affect its reproductive behavior). Because I was relatively new at recording frog calls, I needed a worksheet for all the data I wanted to record, and I had an array of instruments including a thermometer, GPS unit, hygrometer, watch, and altimeter.

While distracted with recording data in the dark, I was caught off guard by what sounded like a very large animal noisily and rapidly approaching. I hastily went through a mental checklist of all the possible critters that would make that kind of racket and had it narrowed down to human, Cassowary, or wild pig. There aren't that many large animals over 10 kg in the forest of New Guinea, much less any that would be noisily traipsing through the forest at night. Isolated for millions of years at the northern edge of the Australian plate as it slowly rotated counterclockwise and approached what is now the equator, the proto-island of New Guinea accreted several smaller island arcs: biogeographic misfits across the vast waterways that divided Laurasia from Gondwanaland millions of years ago. Large placental mammals (outside the recent influx of humans, pigs, and dogs) have never been present, and the larger marsupials mostly went extinct. Mostly.

It was a Papuan forest wallaby that bumped into me as it hopped through the forest, surprising us both, I think. The knee-high marsupial barely paused as it rebounded off my leg and kept right on hopping through the forest. I laughed out loud and scared the poor thing even more. I mean, it was super cute and dopey, and as it hopped away I couldn't help thinking about how and why different lineages of animals succeed and diversify or get outcompeted and go extinct. What was the wallaby doing hopping around at night, bumping into me?! I was giddy from the flood of endorphins and adrenalin but also relieved. The incident was probably funnier to me than it should have been because of the near panic I experienced at something unknown noisily and quickly approaching me in the rain forest in complete darkness. My laughter was self-deprecating and short-lived. I had work to do and so got back to recording the frog call.

Not even two months later, I had settled into a daily routine with my local assistants, people of the Pawaia language group who mostly live in and around the villages of Haia, Wabo, and Soliabedo. Working in PNG meant being a guest of the local people and understanding their tight connection to the land and its biodiversity. While they were training me in the ways of the rain forest, I was training them in standard scientific field sampling methods, which encompassed plots, transects, and identification of organisms we would find. These research methods have now become part of the local communities' array of expertise that enables them to rely on intact biological and cultural systems for their livelihood. We would rise early to conduct 5 × 5 m leaf-litter plots to find frogs, snakes, and lizards (mostly frogs) to establish a baseline for the long-term biodiversity monitoring project that we were setting up in the wildlife management area. Then, at night, we would look for frogs along transects in the forest. Ordinarily, we did four leaf-litter plots each morning and one or two visual encounter surveys after dark. With both day and night activities, the fieldwork was exciting and fun but grueling. The terrain in that part of PNG is mountainous and rugged in a way that I have seldom seen in other places, with incredibly steep slopes that are challenging to hike. Typical days included multiple steam crossings, elevational changes in the 1000 m range, and a dazzling array of biodiversity.

The local people that I employed to help me do fieldwork called the leaf-litter plots "*wok bilong ol pik*," which translates into "pigs' work," an apt description of being on your hands and knees in the muddy forest floor. From the little gems we would find during the leaf-litter plots (e.g., scorpions, centipedes, fungi, fruits, seeds, Cassowary poops, spiders, ants, orchids, ferns, and the focal target organisms), the mornings were the best times to get to know my hosts and coworkers. We enjoyed each other's company tremendously, despite (ok, maybe partially because of) getting filthy every day sorting through the detritus, logs, leaves, and decaying vegetation of the forest. I learned a great deal about the forest from the Pawaia. They told me interesting facts and related stories and myths of spirits and supernatural explanations for some of the things we regularly saw in the forest.

We often found Cassowary droppings during the leaf litter plots. Cassowaries have been called the most dangerous bird and implicated in human deaths and dismemberment. They are big (almost 6 ft tall and easily >120 lb) and have enormous poops full of all kinds of seeds from their frugivorous diet. Old Cassowary poops were fascinating to find, not just because of the sprouting seedlings but also the impressive seed armor (endocarps) in which many of the plants encapsulate their seeds to protect them from either passage through the Cassowaries' digestive system or the rat and bandicoot seed predators. Whatever the driver of the heavy investment to protect the seeds, they were an amazing display of evolutionary adaptation and a source of constant wonder at the merging of form and function.

At night, we used transects to sample frog activity, diversity, and abundance. This typically meant that three local assistants and I would walk a prescribed trail for an hour, looking for frogs; catching, measuring and identifying them; and then releasing them afterward. This work was also a delight, with opportunities for discovery every time we went out. I always tell people that the best time to be out in the forest is at night, when all the cool stuff comes out. On New Guinea, this was especially true. Not only would we find frogs and snakes, but also tarantulas, scorpions, stick insects, geckos, bats, tree kangaroos, and many strange birds like frogmouths and nightjars. We walked slowly at night, each of us looking for frogs on the ground and up into the vegetation. It was at night that we would find the egg-brooding frogs that lay their eggs on the underside of live leaves a few meters above ground. These frogs guard their eggs only at night, rehydrating them before leaving them unprotected during the day. We also discovered one of my favorite behaviors to have ever witnessed: froglet transport. Although the local people knew about this and it had been rudimentarily described, no one had been able to see it enough times to document and measure the behavior.

That all changed in 1995. I was returning from a visual encounter survey in a light drizzle when I saw a strange shiny blob on a fallen branch that had almost crossed the trail. At first, I did not know what I was seeing. I thought it must have been an injured adult frog. Sometimes we would find animals that had

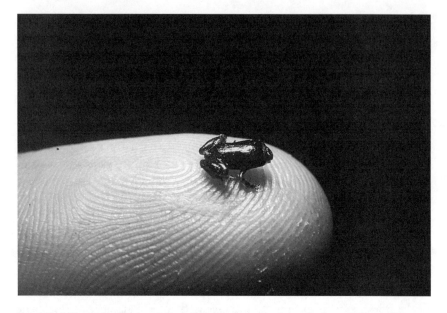

Froglet of Wandolleck's Land Frog (*Liophryne schlaginhaufeni*) on author's finger. Photo by David Bickford.

been injured, escaping a predator but left somewhat disfigured or missing limbs. But this animal was symmetrical and much wider than a normal adult frog. It reminded me of glass sculptures I have seen—smooth, fat, and shiny. Then I realized that it was an adult frog covered in tiny froglets! It was one of my favorite frogs at the site, Wandolleck's Land Frog (*Liophryne schlaginhaufeni*). I stared for a few seconds, thinking "what serendipity!" This was an amazing thing to see in the field, and I felt both amazement and wonder about how such a behavior could have evolved. I could not contain my excitement and had to let my assistants, Lamec, Peter, and Selau, know what we had stumbled across. They caught my enthusiasm and after a brief whispered conversation about what to do next, I asked them to stay with the frog, not to disturb it, but track it if it moved. I would run back to camp to get my camera. When I returned, the frog had moved only a few centimeters, and I was able to count the number of froglets and take many high-quality pictures. We even noted clear azure punctations across the dorsal surface of the froglets—something distinct from the adults. I was psyched; this was a major discovery that I would be able to study in great detail.

That chance observation enabled us to set up a longer-term study with detailed methods to collect data on the novel behavior, which we ended up documenting in three species at that site. It was to be the basis of my first scientific publication and would turn out to be an important contribution to the understanding of not just reproduction but of parental care behaviors in frogs. It is these kinds of discoveries that make every opportunity to be in the field exciting and that keep field biologists on our toes. You never know what you might find.

About the Author

David Bickford is an evolutionary ecologist who has worked in 14 countries as a biodiversity conservation practitioner, field biologist, and university professor. He grew up in Minnesota and developed a strong connection to the natural world through many opportunities camping, backpacking, canoeing, and orienteering with the Boy Scouts of America (mostly in the Boundary Waters Canoe Area Wilderness). David got his PhD in biology at the University of Miami in 2001 and has since published widely on the conservation of biodiversity, anthropogenic drivers of extinction, science communication, and behavioral ecology. He is moving to Colorado with his wife and their kelpie, Datshi.

15

IN SEARCH OF WONDER

How Curiosity Led Me to Madagascar

Phillip Skipwith

I have had the good fortune to participate in herpetological collecting trips in the American Southwest, Mexico, the Dominican Republic, New Caledonia, New Zealand, and Australia. While all these experiences have been memorable, the excursion burned into my mind was my 2018 expedition to Madagascar. At the time I was a postdoc at the American Museum of Natural History (AMNH) working on ecomorphological diversification in Malagasy pseudoxyrhophiid snakes with Dr. Frank Burbrink, head curator of herpetology. I had spent the better part of a year trapped in New York City, producing CT scans of representatives of this diverse assemblage of snakes and combining those data with phylogenomic estimates to quantify how ecology influenced evolution of the skull. Of course, I seized the opportunity to join Dr. Chris Raxworthy on a trip to central-western Madagascar that fall. This was primarily a general collecting trip with an emphasis on pseudoxyrhophiids. After a month of preparation that primarily entailed hemorrhaging money on new field gear, antimalarial medication, and antibiotics I had high hopes would remain unused, I embarked on a 27-hour flight to Madagascar. I'm not sure what I had been expecting, but upon landing in Antananarivo, I was stunned by the press of people in the terminal and in the streets, each peddling rides or advice for dealing with immigration in French. Virtually every street corner was teeming with stalls selling second-hand clothes, rusted car parts, produce, salted meat, and disassembled electronics.

Antananarivo is located in the island's Central Highlands, the breadbasket of the nation. The weather was humid but otherwise mild, and this comfort held while we headed west from the capital in Chris's well-used Land Rover. In addition to Chris and myself, the vehicle hosted our gracious Malagasy driver, two master's students from the Université d'Antananarivo, and Arianna Kuhn, then a doctoral student from my lab at the AMNH. I found myself crammed in the backseat alongside Arianna and one of the Malagasy grad students, wishing there were headrests.

Our first stop would be the small city of Tsiroanomandidy, 130 mi or so to the west. It was a journey that would have taken five hours or less as the crow flies, but it took us two days. Within 50 mi of the capital, the paved roads turned into dirt, and these soon morphed into ravine pocked paths that the Land Rover churned through with mechanical determination. I found myself clinging to Chris's seat with each dip to avoid smashing my face into his headrest and wishing that there were also seatbelts in the back. Every few miles we would pass a group of Malagasy, haplessly staring at a vehicle stalled in a 3-ft-deep trench or struggling in vain to haul it out with raw manpower. All the while my gaze was drawn to cliffs and rolling hills beyond the road, such as it was. This was the first indication I saw of why the capital seemed so depauperate in wildlife. These foothills had been savaged by a combination of agriculture and poorly regulated forestry practices. Scrubs and denuded hillsides stretched as far as the eye could see. We would occasionally pass a small village in a valley, and I found my thoughts drifting toward the morbid. How long would it be before a flashflood would tear through these heavily eroded slopes and bury its inhabitants?

This trip coincided with a presidential election that seemed to have thrown the country in disarray. We found ourselves under the constant reminder that much of our travels would land us in regions where we'd be prime targets for gangs looking for easy victims. After checking into our hotel rooms, we stopped at a bustling bus port, looking for a westbound convoy to join in the morning. Convoys had become commonplace, relying on strength in numbers to dissuade potential attackers in remote areas. Arianna and I waited off to the side while Chris and our Malagasy friends spoke in French to some police officers. Scanning the crowd, I spotted a little boy, perhaps 18 months old, waddling down the dusty road unattended. I was taken aback on a second glance. His legs were bowed outward, the tibia slightly arched. Rickets. He wasn't the only one: a slightly older child with the same condition stood behind him. Evidence of malnourishment and the general lack of resources in this region was presented to us at every turn. Tsiroanomandidy was not a large city, but some streets were easily as densely populated as the markets in Antananarivo. However, there was barely a tree in sight, being limited to a few hostel lawns. Virtually every tree in the city and the surrounding hills had been slain and carried off for timber and firewood. The result was a baking dustbowl with little shade.

As darkness fell, it became clear that the convoy was not leaving until the morning. Luckily, the grounds of our hotel hosted some of the only large trees near the city center. I perused the grounds after dusk in the naive hopes of spotting a leaf-tailed gecko (*Uroplatus* sp.) with my headlamp. Geckos can often be difficult to spot in the tropics. Their eyeshine is comparatively weaker than a mammal's, and dense vegetation does not help in the least. Luckily, there were only five trees on the property, and I instead relied on the ghostly pale cast of most gecko skins at night. In the lowest boughs of an aging tree, I caught the stark white form of something massive on an overhang branch. With its turret-like eyes sealed shut, I found myself staring at a sleeping male Oustalet's Chameleon (*Furcifer oustaleti*), one of the largest chameleons in the world. This individual was easily 2 ft long, including the tail. With some effort, I pried his mitten-like hands from the branch, ignoring his lazy open-mouthed threat display. The next morning, Arianna and I found two additional males of the same species.

We returned to the bus port hoping to join a convoy, which, again, never formed. Instead, we were informed that we needed to hire soldiers as armed guards during our trip. After negotiating with the local garrison commander, Chris was able to enlist two young Malagasy soldiers. Armed with AK-47s, the pair couldn't have been more than 20 years old. Our journey resumed with our armed aides crammed on the benches behind the backseat, assault rifles cradled between their legs. The Central Highlands had been dotted with rice fields and well-provisioned villages. As we descended into the western foothills, this pattern vanished. The cooler weather dissipated and morphed into swampy humidity and oppressive heat. Somehow, there were even fewer trees in the lowlands. The few stands that remained close to the dirt road were but distant memories, having long been reduced to charred skeletons by the locals. Where the higher elevation settlements had hosted brick homes, these lower villages were composed almost entirely of clay. This region supported little agriculture, its residents relying heavily on the relative prosperity of the high-elevation rice fields. People huddled in what little shade they could find, and malnourished children were a more common sight in these parts. We passed several villages that had clearly been overtaken by an unmanaged burn that had gotten out of control, leaving a smattering of charred clay mounds.

We were occasionally reminded of the ever-present risk of bandits upon passing burnt out van chassis, riddled with bullet holes. Our next stop would be the Beanka Protected Area in the Melaky Region. About a day out from our destination, we stopped at a dense stand of trees, a relatively rare sight at this point. As I wandered off a short distance for some privacy, a sharp crack rang out through the trees. I hit the ground, heart pounding, immediately recognizing the sound of a gunshot. Moving back to the car at a crouch, I watched as one of our guards fired, seemingly at random, over the forest. A quick word with Chris revealed that he was warning any bandits hoping for an easy ambush that we were well armed. We moved on, reaching Beanka by nightfall. By this time, I had become accustomed to the wild and sometimes vivid dreams brought on by antimalarial medication. What was more worrisome was the growing gastrointestinal distress that was sapping my strength and ruining my appetite. I chalked it up as a side-effect of the meds.

Beanka was beautiful, the first rich forest I had seen on the trip. It hosted a ranger's station of sorts, a small village of locals who relied on the land, and an extensive karst system. For a week, we struck out at night, led by a local guide, avoiding the blistering heat of day. The first night we found several dwarf chameleons (*Brookesia* sp.) sleeping in the undergrowth. These tiny creatures, shorter than the length of my thumb, would promptly curl into a ball and fall from their perch if startled—an amusing defense strategy that was effectively useless against a team of seasoned field biologists. Each night we found fish-scaled geckos (*Geckolepsis* sp.) scrambling over jagged karst and gnarled tree roots. Numerous gecko genera, such as the Australasian *Gehyra*, have evolved extra fragile skin that can be shed upon capture by a predator and regenerated after successful evasion. *Geckolepsis* proved challenging to catch, their agility matched only by the severity of de-gloving they could survive. After successfully bagging two or three slimy, naked geckos with sticky, scale-covered hands, I opted to avoid these animals for the remainder of our time at Beanka.

The true prizes of our nocturnal excursions were perhaps members of Madagascar's most notable gecko lineage, *Uroplatus*. Arianna and I found a couple diminutive Spearpoint Leaf-Tailed Geckos (*U. ebenaui*) in low shrubs, scarcely larger than a *Brookesia*. Slow and passive, they were poor preparation for what awaited us when one of the Malagasy students gave an excited holler down the trail. We scrambled over the pitted limestone path toward the commotion, narrowly avoiding twisted ankles. All the while, my guts were writhing in agony, a sensation that had been intensifying with each day. When we reached the group, I followed the gaze of the senior grad student. Sitting head up on a thorny tree was a massive gecko, its mottled skin blending in perfectly with the bark—a female Henkel's Leaf-Tailed Gecko (*U. henkeli*), a behemoth by gecko standards that was at least the length of my forearm. She was not alone. Perched on saplings to either side were two larger males, frozen in our lamplight. At night they looked like pale ghosts, but it was evident how impossible they'd be to spot in the cold light of day. These males had been courting the female before being interrupted by our search efforts. I pried one off the tree, reveling in the soft, velvety skin typical of geckos and taken aback by the animal's remarkable strength as it sought to twist around and sink some of its 200 or more teeth into my wrist. It then stared into my eyes and gave a shrill, horrid scream. Our trail guide from the village, a man who had hacked a hapless mouse lemur (*Microcebus* sp.) to death as an offering to our team just moments before, backed away with a warding hand. The larger *Uroplatus* are seen as supernatural creatures of ill omen in some parts of Madagascar, a trend I had seen applied to geckos by the indigenous peoples of Australia and New Caledonia. Indeed, they are uncanny creatures in the wrong light.

Each night at Beanka revealed additional denizens of the karst forest. Several Nosy Komba Ground Boas (*Sanzinia volontany*) were recovered from craters in the karst or under the buttressed roots of trees, waiting for some unwitting mammal to wander by. These were beautiful boas, the largest individual measuring just under 4 ft in length. We even found one female with striking rusty-red and black patterning, which contrasted sharply with the more common mottled green morphs we had been finding. Both Arianna and I managed to receive defensive bites from the comparatively small but elegant pseudoxyrhophiid Cat-Eyed Snake (*Lycodryas pseudogranuliceps*). While venomous, their bites proved very mild, eliciting some minor burning and swelling. These rear-fanged snakes (opisthoglyphous) crept silently through the low vegetation, hoping to surprise a gecko or sleeping iguanian. Upon glancing up, we were occasionally graced by the curious, glowing gaze of a mouse lemur or dwarf lemur (*Cheirogaleus* sp.) as they hunted for insects and small vertebrates.

The days were marred by some of the most relentless heat I had ever experienced. The pain racing through my innards had grown to the point that I could no longer dismiss it as a side-effect of my antimalarial meds. I was losing weight, nearly 14 lb in three weeks, and I could barely hold any food down. Each foray into the forest took all my strength, and it became clear that Arianna was suffering from similar symptoms. To make matters worse, we were running low on clean drinking water, forcing me to use iodine tablets and filters to decontaminate well water. By the time I decided to start a full course of oral antibiotic, I had a full-blown fever topped with severe muscle cramps and chills. On our last day at Beanka, I stood and watched the team prepare our latest specimens in the shade, gripping my stomach, unable to speak through the pain.

Our time at Beanka was but the first portion of a month-long trip, and the sickness that I now knew to be a parasite wreaking havoc on my intestines, the dire need for clean water, and the countless scrapes would fade with time. Our soldiers would eventually be swapped for armed police, all as protection against the bandits we never saw. Somehow, we managed to miss violence altogether. The only evidence we witnessed were more burnt-out cars and, on one occasion, a French tourist who had been beaten and robbed on the road behind us.

Our trip was defined by encounters with dozens of Madagascar's endemic beauties, so in need of protection. I even found myself being chased by the island's largest mammalian carnivore, the fossa (*Cryptoprocta ferox*) while trying to capture its likeness. When I tell tales to laypeople, or anyone else for that matter, I often receive the same responses. Looks of bafflement. Or impatient questions on why I would subject myself to such discomfort or how this path could be lucrative. On rare occasions, someone with little to no interest in nature will shock me by saying, "It is so cool that this is your job." Such sentiment gives me hope. While I can honestly say that fieldwork is rarely easy or even comfortable, it always leaves me with a sense of awe. If I can impart just a fraction of that through stories like this or my photography, then I can take solace in knowing others will see the natural world for what it is: a place of wonder. To witness the

Phillip Skipwith holding a *Phisalixella variabilis* at Beanka. This is a slender, arboreal species closely related to the iconic cat-eyed snakes of *Lycodryas*. This particular specimen was caught at night in a low shrub, possibly stalking sleeping chameleons. This genus is one of four arboreal "cat-eyed" pseudoxyrhophiid snake genera that arose during this clade's radiation during the Cenozoic. Photo courtesy of Arianna Kuhn.

things that we have seen as biologists and natural historians can only leave one with the reverence the natural world deserves.

About the Author

Phillip "Skip" Skipwith majored in biology at the Richard Stockton College of New Jersey (now Stockton University), graduating in 2008. From 2007 to 2009 he worked as a research assistant at the Wetlands Institute in Stone Harbor, New Jersey, assisting in conservation projects. Skip would go on to complete a MS at Villanova University in Pennsylvania with Drs. Aaron Bauer and Todd Jackman in 2011 and a PhD at the University of California–Berkeley with Dr. Jimmy McGuire in 2017, studying the evolution of geckos. From 2017 to 2020, he worked as a postdoc at the American Museum of Natural History with Dr. Frank Burbrink before starting as an assistant professor at the University of Kentucky in 2020.

Part III

FASCINATION AND LOVE FOR THE ANIMALS

16

NEVER WORK ON A SPECIES THAT IS SMARTER THAN YOU ARE

Rick Shine

I adore reptiles, and I always have—ever since my childhood, when I encountered small lizards in the backyard of our suburban house in Brisbane, in subtropical Australia. And I respect reptiles. They are sophisticated and elegant animals, pursuing mysterious lives. Their ecological footprint is vastly lower than that of energy-squandering warm-blooded mammals and birds. So in a very real sense, I'm a reptile zealot. I have devoted years of effort into public relations for reptiles, especially snakes, to convince the general public that scaly creatures have real value.

But intelligence? It pains me to admit it, but by and large, reptiles are not the sharpest tools in the toolshed. Brains are expensive, chewing up more energy per gram than any other type of bodily tissue. So for an energy miser like a reptile, investing resources into cognitive ability is unlikely to be worthwhile. For the average lizard, being smarter doesn't pay off in the cold hard calculus of natural selection. I don't doubt that there are some cold-blooded geniuses out there . . . for example, recent research on lizards has revealed fascinating complexities in social systems and some spectacular examples of cognitive gymnastics. But if you're looking for Animal Einsteins, you probably won't spend much time in the reptile house of the local zoo.

To be brutally frank, my adoration for reptiles is based on many characteristics of these fantastic beasts . . . but problem-solving ability isn't high on that list. Of course, I may be wrong. Humans tend to rate other creatures as "intelligent" when we look into their eyes and perceive some subtle cues that hint at mental activity going on in their brains. And inconveniently, most lizards and snakes are so small that peering into their eyes requires more visual acuity than I can muster.

But a few species of reptiles are big enough for a human being to peer into their eyes. And fortunately for me, Australia is home to some of the biggest lizards on planet Earth. The really outsized species all belong to a single family—the Varanidae—which are known as "monitor lizards" in most parts of the world but are called "goannas" in the Land Down Under. Some of these animals are genuinely huge, especially the Komodo Dragon (*Varanus komodoensis*), which evolved in Australia but is now found only on a few small islands to the north. A large male Komodo can weigh more than 90 kg (200 lb) and can take down prey as large as adult water buffalos. Anyone who believes in the innate superiority of the human race over lowly cold-blooded creatures should take a trip to the island of Komodo, where lizards are in charge of the ecosystem. Komodo Dragons are more than big enough for a person to look them in the eye. And when you do, you realize that the lizard regards you as a potential dinner rather than as a threat to its existence.

The varanid lizards that are still found in Australia are smaller, with the Komodo Dragon's close relative, the Lace Monitor (*Varanus varius*), weighing in at up to 14 kg (30 lb). But that's still a lot of lizard, especially when it stands up on its hind legs, hisses like a steam train, and lunges at you. Goannamosity incarnate. More than 2 m (6 ft) long, with massive claws, serrated teeth, and a muscular tail that can be thrashed around as an additional weapon. I well remember the first big Lace Monitor I ever caught, when I was about 10 years old. Aware of the damage that could be inflicted by those teeth and claws, I leapt upon the lizard with one hand around its neck and the other on its hindquarters, thereby taking those formidable weapons out of play. But I didn't have enough hands to control the tail as well, and a mighty slash of the tail across my face left me with a profusely bleeding nose for the next 30 minutes. By the time I arrived home, my shirt was soaked in blood; my mother thought I'd been hit by a car.

But brute strength isn't the only ammunition in the goanna's arsenal. They are smart. There's something special in the way that a goanna looks at you. They seem to feel that the biblical injunction about humans "having dominion over all the wild animals of the earth" should have been followed by an asterisk that directs you to a footnote saying "except varanid lizards." It's difficult to feel superior when you come back to your tent after a hike to find the side of the tent slashed open, a steamy pile of lizard poo on your sleeping bag, and a giant lizard slumbering on a tree limb several meters above the remains of your camping gear. Especially when the swollen belly of that lizard contains the steak that you had brought along for the night's dinner.

Any herpetologically obsessed child has a few misadventures while they learn how to catch reptiles. That's especially true in Australia, where most of the snakes are venomous. But despite many near misses from tackling deadly snakes, only one youthful encounter inflicted a wound still evident on my now-elderly body: a long scar on my forearm. And that wound was donated by a goanna, not a snake. I had climbed a eucalypt tree in pursuit of a huge Lace Monitor. Perhaps surprisingly, I had managed to seize the lizard without losing any blood—but then was faced with the problem of getting back down. Climbing required at least one hand, meaning that I could restrain the goanna only around the neck. And as soon as I released its tail base, those powerful hindlimbs swung around and the lizard sunk its hind claws into the meat of my forearm. By the time I reached the ground, that laceration was down to the bone.

Okay, you might think, a goanna can defeat a child—but surely an adult human being can stand up to any lizard? I can only wish that were true. As a professional biologist, some of my most humbling experiences in the field have involved showdowns with goannas.

Early in my career, fate put me on a collision course with the Yellow-Spotted Monitor (*Varanus panoptes*) of tropical Australia, about the same size as a Lace Monitor. These giant lizards are the apex predators on tropical floodplains all across Australia's "Top End," and are important traditional food ("bush tucker") for many Aboriginal communities. So when the Australian government approved a uranium mine slap-bang in the middle of the wet–dry tropics of the Northern Territory, the management authorities realized that we needed to know more about the ecologically and culturally significant goannas of the area. In 1982, I was given a contract to do that research. And for completeness, I was asked to look at the ecology of two other species as well: the semi-aquatic "water monitors" (*Varanus mertensi* and *V. mitchelli*).

I was still very early in my career—I had just landed my first academic job—and everything I knew about tropical biology could have been written on a single rosette on the flanks of a Yellow-Spotted Monitor. Nobody else knew much about the tropical fauna, either, so there were no wise colleagues to advise me. I used my contract money to hire my first research assistant—Rob Lambeck, an undergraduate from one of the Uni classes I was teaching—and to pay for our flights from Sydney to the mining town of Jabiru, east of Darwin.

The first morning after our arrival, Rob and I took a four-wheel-drive vehicle out onto the majestic floodplain of Magela Creek. To my uneducated eyes, it was like landing on a different planet—the extraordinary lushness of the wetlands, thousands of waterbirds, massive Saltwater Crocodiles sinking into the muddy water as we approached. And for a keen herpetologist, it was an intoxicating introduction to creatures I had read about in books but never seen in the wild—spectacular Frillneck Lizards (*Chlamydosaurus kingii*), huge Olive Pythons (*Liasis olivaceus*), and venomous snakes (King Browns, *Pseudechis australis*) larger than any serpent in southern Australia.

The goannas were there, too, but they were secretive. We eventually worked out how to find water monitors, by paddling a canoe down pandanus-fringed billabongs or carefully examining creek-crossing sites along the roads. The larger Yellow-Spotted Monitors were more widely distributed, but we could find them on the open floodplain close to the water's edge.

Finding a goanna, though, isn't the same as catching it. Time and again, we found goannas. And time and again, we crept up to them. And time and again, they hurtled off as we approached, into inaccessible sites like crocodile-infested billabongs and massive brush piles. It was infuriating—the lizards knew exactly how close they could allow us to approach before they moved, so on many occasions we were only a heartbeat away from launching a dive at the reptile before it evaporated in front of us.

We caught a few goannas on that first trip, mostly through luck and using 4-m (12-ft) fiberglass fishing poles with nylon nooses on the end. It was expensive work, though, because a noosed goanna would often tear off into cover, snapping the pole around an inconveniently located log. Our best success was with large male Yellow-Spotted Monitors that were courting females when we saw them. The female disappeared long before we could approach, but the romantically aroused males paid less attention to us and often were still searching for their lost loves when we slipped the noose around their necks.

On our second trip, we were better prepared. We constructed goanna traps, 2-m (6-ft) lengths of pipe with a trapdoor at one end, attached to a trigger holding smelly bait such as long-dead fish. Whenever we saw a goanna basking, but failed to catch it, we deployed one of the traps. And mostly, when we returned, the goanna was still there, but now sitting on our trap rather than secured inside it. Captures were few and far between.

Our sample sizes gradually improved, but it was a trying time. We had learned how to find goannas—lots of goannas—but the lizards outsmarted us at least 90% of the time. It was humbling.

Our breakthrough in varanid catching came when I finally recognized a paradox. First, goannas were almost impossible to catch by any of the methods that we tried. Second, Aboriginal people have been catching and eating goannas for tens of thousands of years. So, they must know how to catch these fiendishly clever lizards.

I employed a local Aboriginal man, Uncle Billy, to teach us how to do it. And it was amazing. The next time we found a large Yellow-Spotted Monitor out on the floodplain, I brought the car to a stop. Instead of hurtling toward the lizard as Rob and I had been doing, Billy just got out of the car, then stood still. And after about 10 minutes, he took a step toward the lizard, then stopped again. It was a very slow, very careful approach, with each foot placed down gently. As Billy grew closer to the animal, I expected the lizard to race away—but instead, it slowly sank to the ground, almost disappearing into the grass. And it stayed there as Billy kept up his snail-paced advance, until he eventually simply bent down and took hold of the animal. Even then, it hardly moved.

The lesson, of course, was that a goanna faced with a slow-motion approach may rely on concealing itself rather than fleeing. So, all we needed to do was approach slowly. But my confidence in solving the problem was short-lived. Neither Rob nor I had the awesome patience required for the slow-motion method, especially when clouds of biting flies descended on us. We moved too quickly, and many lizards retreated rather than awaiting capture. Still, our success rates improved, and we gathered enough data to answer the questions that we were asking.

But you have a lot of time to think as you creep slowly, step by step, toward a large lizard that may or may not wait for you to arrive before hurtling away like

The author's nemesis, a Yellow-Spotted Monitor (*Varanus panoptes*), gazing at the camera as it works out new and even more devious ways to make an eminent herpetologist look like an idiot. Photo courtesy of Ruchira Somaweera.

a turbo-charged missile. Lots of time to reflect on the advantages of studying stupid reptiles instead of smart ones. There are thousands of species of reptiles in the world, and the varanid lizards are the only ones that sneer at me. I'd prefer a relationship with animals I can outsmart. I vowed that never again would I conduct a field project on goannas.

About the Author

Rick Shine is an Australian biologist with a lifetime passion for reptiles. He was based at the University of Sydney for many years but recently transferred to Macquarie University. He has studied snakes in many parts of the world, mostly for insights into their ecology and behavior, and has published over a thousand papers during his career. He has won many awards for his research, both nationally (e.g., Prime Minister's Prize for Science) and internationally (e.g., Henry S. Fitch Award). In recent years he has extended his research to amphibians (Cane Toads) and to snakes in the water (sea snakes) as well as on land. But wherever possible, he restricts his studies to reptiles that are as slow witted as the research team.

17

THE REALITY OF GIANT GECKOS

Aaron M. Bauer

I grew up reading *National Geographic*, which in those days ran advertisements in the back for international airlines and freighter ships on which one could book passage. I wrote away to them all and received a lot of mail for an eight-year-old. Most were brochures, but some included novelties, like plastic tikis the color of jade from Air New Zealand. Having already decided that I was going to be a herpetologist (a decision cemented in stone by the capture of a Smooth Green Snake at the age of five) I envisioned myself in the tropics catching reptiles; I had not yet reached the stage of considering why I would be catching them or what I would be doing with them. In 10th grade I read *Mutiny on the Bounty* in class, and this kicked off months of delving into eighteenth- and nineteenth-century literature of exploration in the South Pacific.

Years later, when the time came to pick a project for my doctoral research, I looked for a way to combine my interest in lizards, and geckos in particular, with something that would necessitate travel to the South Pacific. I found what I was looking for in a 1964 paper by the great German herpetologist Robert Mertens. It provided information on the morphology and behavior of captive New Caledonian Giant Geckos (*Rhacodactylus leachianus*). These lizards—the largest geckos in the world—had distinctive short but prehensile tails, and despite their massive size (for a gecko), they were virtually unknown. I discovered that the most recent extensive herpetological field work in New Caledonia had been undertaken by the Swiss zoologists Jean Roux and Fritz Sarasin in 1911–1912 and that little had been done since. In the year that followed, I immersed myself in what literature there was on *Rhacodactylus* and its relatives and familiarized myself with Roux's sepia-toned photos of exotic sounding localities like Oubatche, Bopope, the Ngoï Valley, and the Roches de Hienghène. I also wrote away, largely unsuccessfully, to determine if permission to collect reptiles in this French territory needed to come from Paris or from the territorial capital of Nouméa. In the end, no one responded and, armed only with the name of a potential contact in the largest nickel mining company on the island—a possible source of access to land on which I could collect—I winged my way into the great unknown.

My path to New Caledonia took me the long way around the world. I left California for two months of museum work in Europe and then, after a short stay in Thailand (just long enough to get malaria), I went to Australia. There I visited major museums and carried out fieldwork in South Australia, Western Australia, and the Northern Territory. Although the distance was no greater then than now, in the pre-cell-phone, pre-internet era, home certainly felt much farther away, and one had to consider the implications of working remotely or working alone in the field, or both, when there was no way to call for assistance. Fortunately, I was young and stupid and, like most 22-year-olds, immortal (in my own mind); thus, the lack of communication posed no impediment to my research. I visited spectacular landscapes, made lifelong friends, saw my first agamids, varanids, and elapids and caught my first geckos and pygopods. I also managed to wreck a rental car, necessitating a 20-mi walk to the nearest town, and got lost while collecting at night in the stone country of Kakadu National Park.

After crossing the Coral Sea and the outer reefs of New Caledonia I landed at the international airport in Tontouta, which still had a bit of the feel of a French outpost in the Pacific, complete with patrolling gendarmes. Then, as now, most of the tourist infrastructure was centered on Nouméa, but the capital retained vestiges of its past, like the Place des Cocotiers in the town center and the offices of the South Pacific Commission (now Pacific Community), which at the time occupied a former World War II American military base near the beach at Anse Vata. I paid my respects at the nickel company where my contact, who had never responded to my letters, had apparently long before retired. Thankfully, his replacement, despite knowing nothing about me or my research, arranged a car for me and put me in touch with the permit authorities so I could begin to collect specimens across the island.

Leaving Nouméa, I found the landscape became decidedly more rural going up the west coast, but it was largely agricultural. However, always to the east, the mountains of the Chaîne Centrale beckoned with the promise of forests filled with geckos. Although roads on the west coast were mostly paved, those crossing the mountains were only partly so, and on the east coast, dominated mostly by scattered Kanak (local Melanesian people) villages, even the main road was unpaved outside the few larger towns. In the far northeast of the island the road ran out at the mouth of the wide Ouiaème River, and a small ferry was (and remains) the only means of reaching the foot of the Panié Massif—at 1628 m the highest peak on the island and a well-known site of both floral and faunal endemism.

Driving around the island I soon discovered that my car overheated on any incline, necessitating frequent stops, which provided (or forced) opportunities to collect. I quickly got the knack of catching the terrestrial skinks and the smaller geckos in their daytime retreats. With the reptile fauna nearly entirely endemic, every species was new to me and fascinating in its own right. The skinks, in particular, had not been revised in decades, and most were allocated to one of a few large "trash bin" genera for convenience, but their real affinities were unknown. With no field guides to the island's fauna, I worked mostly from Roux's 1913 monograph and from notes provided by Ross Sadlier at the Australian Museum who, like me, had seen New Caledonia as a fruitful area of study and had traveled there a few years earlier to begin a study on skinks. I had not been successful with the giant geckos, however. Although my headlamp had seemed to work all right in open habitats in Australia, I soon discovered that for night work it was not up to the task of eye-shining in the forest. My nocturnal take was mostly limited to the smaller geckos of the genus *Bavayia*, which were often active on low perches or tree trunks. After several weeks, as my time in New Caledonia was ticking away, I had yet to find the giant geckos I was after. In the south of the island, I was able to get Gargoyle Geckos (*Rhacodactylus auriculatus*), now popular in the pet trade, but then a species that few herpetologists had seen alive. These lizards, with their knobbed skull ornamentation, were often active on low perches and, as I was to discover some years later, were unique in that they regularly fed on flowers, tree sap, and other lizards, in addition to the staple gecko diet of arthropods.

Eventually I heard of someone in a tiny village in the Vallée d'Amoa in the northeast of the island who knew the truly giant geckos, which are locally referred to as caméléons and feared because they are wrongly believed to be venomous. I arrived at my contact's house just as darkness fell, and he told me he had seen them recently on a particular tree. Despite my inadequate headlamp I hoped that knowing precisely where to look would increase my odds. I still could not make out the red eyeshine of these massive geckos, but I was able to make out a body outline at the edge of my light. Although I could see only a bit of the animal, it was enough to confirm its identity. I broke off a long thin branch from a nearby tree and tried to get the end of it above the gecko, which was pointed upward on the tree trunk and had begun to move upward when disturbed by the light. At about 15 ft above my head it was too far for me to reach, so I tried a "wrist rocket" slingshot that I had with me. After a few misses in the dark I was able to hit the trunk just above the gecko, which stopped its upward climb. By shining the light above the gecko, it was induced to move downward, and when it got within range of my branch I flicked it off the trunk and caught it as it fell. At about 22 cm in head and body length it was nowhere near a record size, but it was larger than any other gecko I have ever seen and, indeed, bigger than any other living species of gecko. Its abdomen was soft and fleshy, but the animal as a whole was very muscular and powerful and it let out a loud, deep guttural distress call typical of the species. I could feel both its huge toepads and small but needle-sharp claws hanging on to me and could see the adhesive pad on the tail that Mertens had written about and that had first piqued my interest. In that instant *Rhacodactylus leachianus* became "real" to me.

Although over the years I have worked on species I have not seen or collected in the wild, I feel I never really know an animal unless I have seen it in the field. Even a fleeting glimpse of a species suffices to move it from an almost abstract concept, represented by measurements or characters, to something imbued with a gestalt that renders it more than the sum of its parts. Seeing that first giant gecko moving in the periphery of my light not only fixed a search image in my mind but gave me insight into how this species lives and what its ecology and behavior might be like. I have relived that moment hundreds of times with every

Rhacodactylus leachianus, the giant gecko of New Caledonia. Photo by A. M. Bauer.

new reptile or amphibian species I have encountered in the field. Although I may not have a precise recollection of every locality where I found a given species, I do retain an almost photographic-quality memory of its immediate surroundings, its posture, its movement, and the minutiae of its appearance. For places like New Caledonia or southern Africa, where I have been on dozens of field trips, the sum of these memories has, over a period of decades, contributed to a more comprehensive "feel" for how entire herpetofaunas function and how they may have evolved.

But that initial impression, of necessity, gave way to the practical need of securing my captive, which I transferred to a cloth bag that I tied with the tightest knot I could. Fresh off my victory, within minutes I found another, though smaller, gecko closer to head height on a trunk. I recognized it as another species of giant gecko, *Rhacodactylus* (now *Mniarogekko*) *chahoua*. If anything, this was even more exciting, as this species had been described in 1869 and not seen again until more than 100 years later. It, and a second specimen collected on an adjacent tree, became the nucleus of my first research paper in which I designated it as the neotype of the species, replacing the original name-bearing holotype, which had gone missing sometime after 1883. Although it had taken me until the end of my stay to finally collect these giant geckos, I returned to Nouméa satisfied with the trip and ready to ship the lizards back to the United States. The night before packing the lizards for their long flight (I still had a month of fieldwork in New Zealand before heading home), I discovered that the bag containing my only *Rhacodactylus leachianus* was empty. I looked everywhere without success and went to sleep thoroughly defeated, a major aim of my trip unfulfilled due to my carelessness. But, snatching victory from the jaws of defeat, the next morning a final search revealed that the gecko had managed to get into a rolling window shutter, apparently by lifting an impossibly heavy access panel with its head.

Armed with a far better headlamp and the knowledge gained on that first trip, I returned to New Caledonia five more times during my doctoral studies and a dozen more after that. On every trip, "lost" species were rediscovered or new species were found, and for each of these there is a stored snapshot in my brain of that first view of something new. These images recall that visceral sensation of discovery and fuel my enthusiasm for the next field trip, and the next, and the next.

About the Author

Aaron M. Bauer was born in Manhattan but prefers to think of himself as a citizen of the world. He obtained his PhD at the University of California–Berkeley in 1986 and since 1988 has been at Villanova University, where he holds the Gerald M. Lemole Chair in Integrative Biology. His research interests focus on the systematics, biogeography, and evolutionary morphology of squamate reptiles, especially geckos and other lizards of the Southern Hemisphere and the Old World Tropics. His fieldwork regularly takes him to southern Africa, the Indian subcontinent, and Borneo in addition to Australia and the South Pacific. He also maintains a research program in the history of natural history.

18

FOLLOWING THE MOLE (SALAMANDER) TRAIL
A Forty-Year Cross-Country Journey

Susan C. Walls

Standing knee-deep in a temporary woodland pond, I lifted my dipnet from the water to find an odd creature with smooth skin, four legs, and big floofy, feathery, bright-red external gills wriggling in decaying leaves and muck in the net's bottom. I had never seen such an animal before: I was bamboozled by a larval Mole Salamander (*Ambystoma talpoideum*). For me, this encounter was what James Joyce and Joseph Campbell called "aesthetic arrest": the larva's gill arches and body were painted with gold flecks of shiny iridophores. The highly vascularized, bushy gill filaments—when viewed under a dissecting microscope—circulated oval-shaped red blood cells that moved along arterial pathways like bumper cars. I was hooked.

Thus began what would become a life-long passion for, and research focus on, salamanders—especially mole salamanders of the family Ambystomatidae. Unlike many others, I didn't become enamored with amphibians and reptiles until I was a young adult and enrolled in an undergraduate field ecology course at Mississippi State University. As part of a class field project to study ecological interactions in temporary (ephemeral) woodland pond ecosystems, I sampled a small pond with a dipnet, looking for amphipods and isopods (my assignment in the class). I found plenty of my targets but, more importantly, I found a new passion to study mole salamanders, which would take me to three corners of the continental United States during the course of my career.

Later that year, I returned to the same pond one winter night. The darkness was enshrouded in a cold but gentle rain—perfect salamander weather. As I approached the pond, the water's surface bubbled in the light of my headlamp.

I ran a dipnet through the pond, only to struggle underneath the weight of its contents as I lifted it out of the water. Dark, purplish salamanders poured over the edge of the dipnet. As I sorted through my catch, I realized that I had stumbled upon an en masse breeding aggregation of another species, the Spotted Salamander (*Ambystoma maculatum*). Scores of 10-in salamanders, their purple bodies punctuated with bright orange and yellow spots, looked up at me with dark round eyes and blinked in the bright light of the head lamp. I had never seen such a sight before, nor have I seen it again in the 40 years since.

It turned out that there was yet a third species of *Ambystoma* that used this little pond for breeding. The Marbled Salamander (*A. opacum*) capitalizes on temporary ponds like this throughout its expansive distribution, ranging from New England to northwest Florida, then west to eastern Texas, Missouri, southern Illinois, Indiana, and Ohio. This widespread distribution would afford me the opportunity to work with this species at multiple points in my career as I moved around the country. I focused study on this species, along with Spotted and Jefferson Salamanders (*A. jeffersonianum*), in New York's Hudson Highlands when I became an assistant professor at City College of the City University of New York. Many years later, a discovery of populations close to where I currently live in northern Florida allowed me to rekindle my fondness for Marbled Salamanders and examine how this species is faring at the extreme southern end of its range.

I soon left Mississippi for graduate school in Louisiana where the same three species—Mole, Spotted, and Marbled Salamanders—co-occurred. This set the stage to continue addressing questions I initially posed in my undergraduate days, but there was just one hitch: I had a hard time finding them. I searched the university's museum records for historic localities and found "hot spots" of collections at key locations, in one instance resulting in 300 adult Marbled Salamanders crammed into gallon jars filled with preservative. I thought, "Oh, boy! This must be a great spot for Marbled Salamanders!" A visit to the site exceeded my expectations: the habitat was perfect. The next summer, before the fall salamander breeding season, I hauled left-over plywood boards from construction sites out to the dry pond depressions to provide cover under which gravid females could nest and lay their eggs. But it was not meant to be: I never found a single salamander over the years that I checked the area. I eventually found another field site, but the sad realization sunk in that, most likely, the collector had taken all adult individuals as they arrived at the site to breed. All 300 salamanders in the collection were stockpiled over a short period of time, confirming my suspicion. The worst of it was that no research had ever been done with those animals; the extirpation of that population was in vain.

Following graduate school, I was fortunate to pursue a postdoctoral research position in Oregon. With its mountainous terrain, cool temperatures, lush evergreen forests, and beautiful waterfalls, rivers, and lakes, living in the Pacific Northwest had always been my pipe dream. Plus, species of mole salamanders that were new to me lay in wait! Fieldwork with the Long-Toed Salamander (*A. macrodactylum columbianum*) proved to be one of my biggest adventures yet. My main study site with this species was a high-altitude ephemeral pond south of Sisters, Oregon, nestled in the Cascade mountain range. Like many montane amphibians, this salamander breeds at snowmelt just as soon as the pond's surface is free of ice. In warmer months, one could drive right to the pond; however, in the winter and spring, getting to the pond required skis or snowshoes. Having never skied before, I chose the latter option. The uphill trudge through wet, heavy snow at high elevation proved a challenge for someone acclimated to the swamps of Louisiana. However, the effort paid off when we reached the pond: we found a couple of dark, slender adult salamanders, decorated with gold across their backs, swimming in the ice-cold water. I knew nothing about how this species bred. I found out, however, when I lifted a submerged flat rock and found clusters of freshly laid eggs clinging to its underside.

I continued to visit the pond as the season progressed. Remarkably, hatching of the larval salamanders coincided with a prolific bloom of plankton (copepods and cladocerans)—food for the larvae. As the days grew longer and the water warmed, the plankton population eventually crashed. I noticed that the larval salamanders were very thin and had disproportionately enlarged heads. What could they possibly be eating in this prey-depauperate pond?

I collected some larvae to examine in the laboratory. I sacrificed a "big-headed" morph and discovered that it had significantly enlarged vomerine teeth, just like the known "cannibal morph" in some populations of the Tiger Salamander (*A. tigrinum nebulosum*). Remarkably, after about a month in the lab—reared in isolation from other larvae and fed plentiful prey—a group of wild-collected, big-headed individuals transitioned to more "normal" looking larvae with smaller, rounder heads, fuller bodies and limbs, and reduced vomerine teeth. Based on this morph's similarity to that of *A. t. nebulosum*, I concluded that this population of Long-Toed Salamanders likely developed cannibal morphs in a prey-depauperate environment.

Regrettably, my stint in Oregon came to an end too soon. I moved to New York for my first academic job and more field studies with a trio of *Ambystoma*. But that, too, changed when an opportunity arose to get closer to my family in Mississippi. Thus I moved back to the South—then again (back to Louisiana)—and again! I finally settled in Florida, with the rare opportunity to work with the highly imperiled Frosted Flatwoods Salamander (*A. cingulatum*). This time, however, the motivation was very different, as this species was quickly slipping into oblivion. The focus of my work shifted to conservation. Using drift fences and funnel traps, my team and I began catching adults as they migrated to wetlands to breed, marking and measuring them before releasing them into the pond. This technique would allow us to follow individual growth and survival over time, thus providing an estimate of how well populations were doing. We also began capturing free-swimming hatchling larvae and moving them to water-filled mesocosms to enhance their growth and survival in that predator-free and prey-enriched environment. Once larvae metamorphosed, we marked and released them at their natal pond. Only time will tell if this head-start approach is working.

The panhandle region of Florida is considered a biodiversity "hotspot," yet, with rare exceptions, the region (as well as the rest of the state) lacks ponds supporting multispecies assemblages of *Ambystoma*, as I had experienced elsewhere. Indeed, Spotted Salamanders don't occur in Florida, and Marbled Salamanders—occurring at the extreme southern edge of their range—are only patchily distributed along a few river drainages in the panhandle region. I longed to see a Marbled Salamander again and finally had the opportunity nearly a decade and a half after moving to Florida. One fall I descended into the dry basin of a seasonally flooded swale on the west side of the upper Suwannee River; the basin was ringed with saw palmetto and dotted with decaying logs of varying sizes underneath a canopy of cypress trees. The ground was bare other than a carpet of cypress leaf litter. In late November, the temperature was still unseasonably warm, and it had not rained for many weeks. The habitat was perfect, but would there be any salamanders given the warm and dry conditions? Many logs had to be turned to get an answer to that question, but, eventually, my friends and I turned a log to uncover a pair of female Marbled Salamanders sitting on a communal mass of developing embryos—a double clutch! Repeated visits to different parts of the expansive dry swale revealed only a few more salamanders. Did a viable salamander population still exist here? Were there only a few salamanders because of the dry conditions? Would the winter rains eventually fill the depressions and allow the larvae to hatch?

With so many species experiencing declines and facing the threat of extinction, so-called common species with broad distributions—like the Marbled Salamander—are typically not even on the radar of concerned conservation biologists. However, anecdotal reports of more and more common species are indicating that they, too, are likely declining. Populations at range edges—like the Suwannee River population of *A. opacum*—are especially vulnerable. Empirical studies are beginning to reveal that climate change is leading to both expansions and contractions of species' distributions. As the impacts of climate change

A communal clutch of developing Marbled Salamander embryos, guarded by two females, found underneath a log on the dry basin of a temporary pond in Florida. Photo courtesy of Sam Sweet.

escalate, field guides will undoubtedly have to be revised to reflect shifting distributions of many species.

As I sat by the overturned log admiring the pair of female salamanders and their communal clutch, I debated whether I should roll the log back over and leave them be or whether I should collect the females and their eggs, rear the larvae to metamorphosis myself, then later return them to their natal site. I flipped a mental coin and "collect animals" won. As I later watched the larvae grow and develop, I knew I had made the right choice: the rains never did come, and any eggs left on the landscape perished, resulting in another year of complete reproductive failure for that population. Final score in the game of ecological roulette? Environmental uncertainty one, salamanders, zero.

About the Author

Susan C. Walls is a research wildlife biologist with the US Geological Survey (USGS), Wetland and Aquatic Research Center, in Gainesville, Florida. Susan began her career in academia, with her first position at the City College of the City University of New York followed by a second position at the University of Southern Mississippi. In 2000, she left academia to work with the Amphibian Research and Monitoring Initiative within the USGS. Her research focuses on developing science-based strategies to address complex conservation challenges, particularly with imperiled amphibians. Her primary interest is in pond-breeding salamanders of the genus *Ambystoma*.

19

CHANCE, MYTH, AND THE MOUNTAINS OF WESTERN CHINA

Alan H. Savitzky

In principle, biologists regard the world's varied biota with equanimity, recognizing the equivalent intrinsic value of each species. In reality, however, biologists are like other people—we play favorites, and some species simply capture our interest and imagination more than others. Often these are the same iconic species that nonbiologists find engaging, like the giant panda, the platypus, and even the coelacanth, a fish heralded as a "living fossil" when it was captured by a commercial trawler along the South African coast in 1938. More often, however, biologists bestow their favoritism upon species that have some unusual characteristic of their anatomy, behavior, ecology, or physiology or that occupy a unique position on the evolutionary tree of life. If a species is also rare—or at least rarely encountered by scientists—so much the better. Among reptiles we can see these elements of subconscious favoritism play out in two species nested in the family tree of lizards. At 3 m in length, the Komodo Dragon (*Varanus komodoensis*) is the largest living lizard, and biologists are as enamored of it as are other people. However, a diminutive relative of that species, the Bornean Earless Monitor (*Lanthanotus borneensis*), is even more intriguing to many herpetologists, not only because of its extraordinary rarity but because, in the 1950s, two biologists described its anatomy and suggested it was the closest living relative of snakes. Although additional specimens have come to light in the intervening years, and it is no longer considered a "missing link" between lizards and snakes, the Bornean Earless Monitor has lost little of its fascination for herpetologists, a testament to the lasting impact of a compelling evolutionary story.

In a similar way, a strange viper from the mountains of Southeast Asia occupies a position on the boundary between science and myth, at least among snake biologists. In 1888, Belgian herpetologist George Albert Boulenger, then working at the British Museum (Natural History), was studying amphibians and reptiles collected in Burma by the Italian naturalist Leonardo Fea when he encountered a single specimen that he immediately recognized as unique. Boulenger assigned it to a new species and genus, *Azemiops feae*, now sometimes called Fea's Viper. However, Boulenger initially seemed a bit uncertain as to its evolutionary affinities. He included the description of his new species in the section of his paper dealing with the Elapidae (cobras and their relatives), yet he noted its similarity to the most basal group of vipers, the night adders (*Causus*) of Africa. Complicating Boulenger's interpretation was his belief that the new species' closest relative was in fact a genus and species only recently described from Venezuela, known from a single specimen that had reached the German herpetologist Wilhelm Peters after arriving in Hamburg with a shipment of lumber. Within a few years, however, Boulenger realized that the purported South American locality for Peters' specimen had been incorrect and that Peters' specimen, in fact, belonged to a previously described species of African night adder. In 1890 Boulenger unequivocally included *Azemiops feae* as a basal member of the Viperidae, unusual among vipers in having smooth, rather than keeled, dorsal scales. He also described its un-viperlike appearance, with a narrow head and a color pattern of dark gray with narrow light crossbands that were often broken and offset at the middle of the back.

Despite its uniqueness, Fea's Viper resided in biological obscurity until 1971. Herpetologists George Rabb and Hymen Marx had been studying the evolution of vipers, especially the Old World subfamily Viperinae. The other subfamily, the Crotalinae, is believed to have arisen in eastern Asia and expanded into the New World, where it radiated explosively into such iconic groups as rattlesnakes and bushmasters. Known as pit vipers, the Crotalinae are named for the prominent pit organs in front of the eyes, which are exquisitely sensitive to infrared radiation, the wavelengths of radiant heat that emanate from their largely mammalian prey.

In those pre-genomic days, Marx and Rabb had gathered a large anatomical data set and were exploring new methods to infer evolutionary relationships from the quantitative analysis of morphology. They had identified the most basal members of both subfamilies of vipers, but Fea's Viper was in a class by itself. Known from only a few specimens and limited to the mountainous border region of China and adjacent countries, its external characters were less viper-like than any other member of the family, and its skull was mystifying. Fea's Viper was clearly not a pit viper, since it had no pit organs, but its skull had characteristics of the Crotalinae, not the Viperinae. It was, in some sense, a "pitless pit viper." Marx and Rabb enlisted ichthyologist Karel Liem, a gifted morphologist who had trained under herpetologist Hobart Smith, to study the anatomy and phyloge- netic position of Boulenger's distinctive viper. They placed it in a new subfamily all its own, the Azemiopinae, the first new subfamily of vipers in over a century. Fea's Viper had emerged from obscurity to attain celebrity status among snake biologists. I was about to begin my graduate studies on the morphological sys- tematics of snakes, and the 1971 publication by Liem, Marx, and Rabb was seared into my brain.

So the stage was set: a rare species from a remote region, with an unusual anatomy and a unique position on the snake tree of life. Unlike the Bor- nean Earless Monitor, which proved not to be especially close to the origin of snakes, Fea's Viper has maintained its position as the so-called sister group to the pit vipers and still resides in its own subfamily. Over the years additional specimens came to light, and a few live individuals even made their way to researchers and zoos in the United States. Still more surprisingly, a second species was described in 2013, based on specimens from northern Vietnam. Even so, few Western biologists had encountered Fea's Viper in the wild. I had seen one at the Dallas Zoo years ago, among the first live specimens to reach the United States—fascinating but somehow diminished in its captive cir- cumstances. Still, like a platypus to a mammologist or a condor to an orni- thologist, it occupied a prominent and, it turned out, a readily accessible shelf in my deep memory.

Many years had passed since I thought about Fea's Viper, aside from mention- ing it occasionally in passing when discussing snake diversity in my herpetology class. My research interests had also evolved, from anatomy to conservation and eventually to chemical ecology. I had been working for about 20 years with my colleague and friend, Akira Mori of Kyoto University, on the chemical defenses of a group of Asian snakes, members of the same subfamily that includes the gartersnakes of North America. Nestled within that subfamily, the Natricinae, is *Rhabdophis*, a genus of so-called keelbacks that has evolved specialized struc- tures in the skin that rupture when the snake is attacked, spewing a distasteful, irritating, and toxic fluid that deters the predator. The active ingredients in the fluid are bufadienolides, a class of toxins known primarily from toads, which manufacture them in the warty glands of their skin. The only other animals known to make bufadienolides are fireflies. Akira was very familiar with the sole Japanese member of the genus, the Yamakagashi (*Rhabdophis tigrinus*), which feeds primarily on frogs and toads. He had suggested in the 1980s that perhaps the snakes are capable of storing—or sequestering—bufadienolides from toads they consume as prey. After about a decade of work involving many students and collaborators, we had demonstrated that Akira was right about the source of the Yamakagashi's toxins.

Subsequently, Akira had received grant support to study other species of *Rhabdophis*, and he had assembled a large team of colleagues working across Asia. An early goal was to clarify the evolutionary relationships of the natricines with skin glands, which proved to belong to a single lineage. That study also revealed that a group of species in western China had shifted their primary diet from frogs to earthworms. Surprisingly, the snakes are still toxic, but the worms they consume are not. Something else must be serving as the source of their defensive toxins, and we had traveled to China a few times to try to solve that puzzle. One clue was the odd chemical structure of the worm-eaters' toxins, which included some compounds different from those of toads. Our studies ultimately revealed that the source of the distinctive toxins were firefly larvae, unusual prey for a snake.

Our Chinese collaborators were located in Chengdu, the capital city of Szechuan Province. The Chengdu Institute of Biology is renowned as a center of herpetological research, and from that base we traveled to neighboring regions, searching for species of worm-eating *Rhabdophis*. In September 2018 we were headed southwest to neighboring Yunnan Province for three intensive days of fieldwork in the mountains along China's border with Myanmar. Our goal was to collect Leonard's Keelback (*Rhabdophis leonardi*), a member of the worm-eating group that we had not yet seen. The species had been described in 1927 from a specimen collected a few kilometers across the border in what was then Burma. Our Chinese colleague Li Ding and two of his students had driven ahead with most of our gear, while we awaited a final permit required for international researchers. Akira, his former postdoc Teppei Jono, and I took an evening flight from Chengdu to Tengchong, a city known for pitched battles over control of the border region during World War II. We were met at the airport by a driver and one of the graduate students, Yige Piao, who had asked to be called "Spoon," an English translation of a homophone for his name. We stayed that night at an odd little hotel where the stairway, floors, and bathroom were decorated in laminate proclaiming *The Motorcycle Diaries*, the 2004 film based on Ché Guevara's memoir of his youthful journey across South America.

The next morning we drove west, past endless paddy fields, to Yingjiang, even closer to the Myanmar border. At midday we checked into the upscale and aptly named Xinhua Grand Hotel, from which we could drive to field sites in the mountains above the city. That afternoon we drove up Gongshan Mountain, a forested upland that stretches toward the nearby international border. After scoping out the terrain, still largely populated by Indigenous peoples, we stopped for a dinner of stir-fried dishes cooked behind an immense open-air counter heated from below by six wood-fired ovens. We road-cruised for several hours, finding only a large White-Banded Wolf Snake (*Dinodon septentrionalis*), which looked eerily like one of the species of venomous banded kraits (*Bungarus*). Aside from a few fireflies, which we needed for our chemical studies, we saw little of interest and returned to the hotel, exhausted, well after midnight.

The next morning we headed back up the mountain with "Spoon" and Cheng, the other student, while Ding stayed behind (the vehicle seated only five, so our three Chinese hosts rotated). It was cool and gloomy, with mist settling on the higher elevations as we visited a small sand quarry and, more interestingly, a Buddhist cemetery. Many of the ancient stone crypts appeared to emerge from the hillside like dormers on a pitched roof, their ancient stones supporting lichens, mosses, and ferns. Still, after finding no snakes we returned to town to recharge over a lunch of potstickers, boiled peanuts, and stir-fry.

Later that afternoon we again drove the switchbacks up Gongshan Mountain, revisiting the cemetery, with no luck. We returned to the quarry about 2000 hours, as the light was fading. It was decided that Ding would road-cruise on his own, while Akira, Teppei, Spoon, and I walked along the road to see what we could find at the edge of the dense forest. We spread out along both sides of the road, far enough apart to catch a glimpse from the nearest headlamp. I was walking downhill along the road, mainly staying on the more densely vegetated side from which the slope extended upward, occasionally stepping over the concrete drainage ditch that ran alongside the pavement. Such ditches are ubiquitous on montane roads that cut through the wet forests of eastern Asia. For all the walking, I saw no vertebrates whatever. Those adjacent to me on the road also saw little, although Teppei collected several genera of firefly larvae.

Eventually the person below me turned and trod back up the road, and the ones above me descended until we had all gathered. Akira had been the farthest upslope and, unlike the rest of us, had made two excellent captures, a slug snake, belonging to the complex and bewildering genus *Pareas,* and a Mountain Pitviper (*Ovophis monticola*). We crowded together to examine Akira's catch, admiring both of his snakes when, for no particular reason, I happened to glance to the side. In that moment, the beam of my headlamp landed on a slender, gray body with narrow, widely spaced light crossbands, moving slowly uphill in the drainage ditch. Its head was already under a mat of accumulated vegetation, but I didn't need to see it. It had not occurred to me until that moment that we were within the range of Fea's Viper, but the mental image seared in my memory in college came immediately to my consciousness. With all the alacrity a 68-year-old could muster, I jumped into the ditch and shouted to Spoon to pass me his tongs. With shockingly little effort, I grabbed the snake with the tongs as it continued pushing under the vegetation, then lifted it onto the road, where it was quickly

The rare and strikingly patterned Fea's Viper (*Azemiops feae*) shortly after capture on Gongshan Mountain, Yunnan Province, China, in September 2018. Photo by Alan H. Savitzky.

bagged. In the space of a minute, the evening had been transformed from an uneventful roadside stroll to an encounter with a mythical serpent.

We walked back uphill as a group, continuing to scan the roadsides, as I replayed the short mental tape of the snake's capture over and over. Soon after we had returned to the quarry Ding arrived at the rendezvous point, having found two leaf-green snakes, a Green Cat Snake (*Boiga cyanea*) and a spectacular red-eyed Yingjiang Green Pitviper (*Trimeresurus yingjiangensis*), a species that would not be described until the next year from other specimens collected in the region. We spent over an hour photographing our catch, of which I was naturally most intrigued by the Fea's Viper. Its blue-gray dorsal color and narrow reddish-orange bands made it unmistakable, and its spectacular facial coloration, with bold orange stripes against a gray field and shockingly white eyes, seemed in that moment to be the most beautiful I had seen on any reptile.

We still had not seen a single Leonard's Keelback, but the trip was suddenly a complete success from my perspective. The following evening, our last night on Gongshan Mountain, our luck changed, and we found two subadult Leonard's Keelbacks on the road, the first one alive and the second freshly killed but still in good condition. We were able to get samples of their defensive chemicals from both and stomach contents from the live one, which contained both an earthworm and a firefly larva. We flew back to Chengdu the following afternoon and two days later left.China for Japan.

So much of our lives is governed by chance, and I think of the confluence of unlikely events that put me at that precise point along the road up Gongshan Mountain, admiring Akira's catch, just as the Fea's Viper was passing by. I think of its unhurried movement and the fact that it was constrained within the cement drainage ditch, limiting its options for escape. Most of all, I think of my sidelong glance, for no reason I can recall, and the instant recognition of a color pattern I'd scarcely thought about in years. Snakes of all sorts are fascinating, and many are beautiful, but having the privilege to find one that is also rarely encountered—an object of myth and legend, not to mention desire—alongside the road in the misty borderlands of China was a special gift.

About the Author

Alan H. Savitzky has been a professor of biology at Utah State University since 2011, prior to which he served on the faculty of Old Dominion University in Virginia for 29 years. He received his BA from the University of Colorado and his MA and PhD from the University of Kansas (with his dissertation research supported by a Smithsonian Predoctoral Fellowship). Most of his research has focused on the biology of snakes, especially their morphology, development, ecology, and conservation, with a particular emphasis on the evolution of complex adaptations. For about the past 25 years he has increasingly focused on the sequestered chemical defenses of Asian natricine snakes with his colleague Akira Mori of Kyoto University. He has conducted fieldwork primarily in the United States, Latin America, and eastern Asia.

20

DIVE IN THE AIR BESIDE A RICE PADDY

A Moment to Grab an Eluding Snake

Akira Mori

Since I became a snake researcher nearly 40 years ago, the question I have most frequently been asked is "Why are you studying snakes?" Of course, my answer is "Because I love snakes." Then, the next question is always "Why do you love snakes?" Truly, I have no idea why I came to love snakes. What I can answer is "I was born to love snakes."

I was born in Osaka city. Osaka is the second largest city in Japan, but my hometown was located in a peripheral suburban area, which was just developing as a residential town at that time. Small rice paddy fields were scattered around my house, but there were no nearby woods, river, or swamp where I could easily see reptiles and amphibians. I was not a "naturalist" when I was a kid. Although I often kept goldfish and many kinds of insects, such as horned beetles, grasshoppers, fireflies, and cicadas as pets, I never kept amphibians or reptiles until I was 18 years old. I was usually playing in my house or in front of my house with my friends or with toys rather than going outside to explore the wild. Nonetheless, I liked animals in general and particularly snakes. I had a large collection of small rubber models of "creepy" animals: spiders, centipedes, scorpions, lizards, and, of course, snakes. I collected as many kinds of rubber snakes as possible (not so many were manufactured at that time, though). I liked to watch TV programs about nature and animals, and I spent much time reading my favorite color guidebooks and magazines about animals, especially reptiles, over and over again. When I was in my twenties, I found an old notebook deeply hidden on my shelf. In it, I found my childish pencil drawings of long and undulated creatures.

Apparently, these depict snakes (I could recognize eyes and opened mouths). I guess those are my drawings at the age of three or four.

My late grandmother told me a long time ago that when I was a kid, a small snake once came into our house through the front door and my family panicked. I have no memory of that. However, I do remember that I sometimes heard snake–rat chasing sounds from above the ceiling of my old wooden house. A light, staccato sound made by a running rat (perhaps *Rattus tanezumi*) was followed by a low, slithering sound caused by a snake (most likely a Japanese Ratsnake, *Elaphe climacophora*) chasing the rat.

My first real encounter with a wild snake was probably when I was 9 or 10 years old. As a holiday outdoor activity, my mother had taken me to gather chestnuts in the mountains. When I was alone looking for chestnuts in a shady, small wooded area, I saw a large snake, probably a Japanese Ratsnake, slowly moving on the ground a few meters away from me. The snake apparently did not notice me and continued its slow exploring movements, which I silently watched for a while staying a couple of meters away. After experiencing a short, mysterious, dreamlike time, I quietly left without disturbing the snake. I did not tell my mother or anyone else that I had seen a snake. Somehow, I felt I should not share my first encounter with a wild snake with anyone, instead keeping it as my sacred secret.

The second encounter was my first great snake hunt. When I was in my first year of high school, one of my classmates took me to a rice paddy owned by his family where, he said, there were many snakes. On the mainland of Japan (Honshu), the best place to find snakes is rice paddy fields. Eight species of snakes occur in most areas of Honshu, and all of them can be found in or around paddy fields. The two most common species, Japanese Striped Snake (*E. quadrivirgata*) and Tiger Keelback (*Rhabdophis tigrinus*), are easy to find in paddy fields from spring to early summer. They are either basking in front of shelter, such as stone walls beside the paddy, searching for frogs along foot paths, or swimming in the water-filled paddy. The paddy field that my classmate showed me was a great habitat for snakes. I looked at a stone wall where he had often seen snakes . . . and . . . there they were! I quickly stretched out my hand to catch one snake after another. Those were adult-sized Japanese Striped Snakes, and they were everywhere. I guess I collected more than 10 snakes within several minutes. My classmate was impressed that I was good at catching snakes, even though it was my first time. We brought them back to the high school and kept them in a large glass cage in a room for the biological experimental class (the classmate belonged to a student biology circle at the high school). That night, I asked my family if I could keep the snakes at home, but they said no. I was easily persuaded to buy and keep two small birds, budgerigars, instead of keeping snakes. Meanwhile, the

snakes that were left in the high school room escaped from the cage and disappeared. Neither one of us knew how adept snakes are at escaping from a cage. My
classmate was seriously scolded by a teacher who was in charge of the biology
circle. Fortunately I was not scolded because I did not belong to that circle. He
quit the biology circle after this incident. (I'm sorry about that, my friend.)

My snake-devoted life began when I was an undergraduate student at Kyoto
University. There I joined a biology circle called the Wildlife Research Association. I periodically went to the Ashiu Experimental Forest of Kyoto University,
where I found plenty of snakes. I read many scientific papers written by Prof.
Hajime Fukada, a pioneer in ecological studies of snakes in Japan, to learn field
methods and techniques for studying snakes. Then, I started a basic mark-and-
recapture study in Ashiu by myself and also began to keep snakes in my house.
This time I was clever and skillful enough to persuade my family to allow me to
keep snakes at home.

My snake-catching methods have evolved (and improved) over the years. I have
learned suitable ways of catching snakes, specific to the species, by watching the
behavior of these animals in the field. For example, among the eight species of
snakes on Honshu, six species are active during the day. Among them are the two
most common species, Japanese Striped Snakes and Tiger Keelbacks, which usually
maintain relatively high body temperatures and move swiftly. Whenever I find these
snakes, I have to move quickly to catch them. If a snake notices me approaching,
it will swiftly move into a nearby bush or shelter. To overcome this escape tactic,
I have to dash toward the snake and dive in the air while stretching out my hand as
far as possible to grab the fleeing snake, just like an action movie star running and
diving away from an explosive bomb. It is the most delightful moment when I can
feel the snake's body in my hand while my body is touching down on the ground.

In the field, I often notice the presence of a nearby snake by hearing a specific
sound caused by its slithering rather than by seeing it. In that case, I quickly turn
to the sound source, while looking for the sound maker and stretch out my hand
to prepare for grabbing a presumed snake. On Honshu, the only dangerous venomous snake is a pit viper, *Gloydius blomhoffii* (Mamushi). If I identify the sound
maker as a Mamushi, I immediately abort hunting before my hand reaches the
snake and touches the body. Fortunately, the Mamushi is relatively small and is
generally not very aggressive, and I have never been bitten by a Mamushi while
hunting snakes. However, caution is necessary when I grab the Tiger Keelback,
a rear-fanged natricine. Although this snake is generally calm and does not bite
unless molested, its venom is potentially lethal, and several cases of human death
by its envenomation have been reported.

The above diving hunt is not a suitable way to catch snakes in areas other
than Honshu. On Okinawa Island of the Ryukyu Archipelago, which is located

in the subtropical region of Japan, much more caution is necessary because a large pit viper, the Habu (*Protobothrops flavoviridis*), is common. Similarly, in tropical regions such as Taiwan and Borneo, a quickly escaping snake might be a dangerous cobra.

In this regard, Madagascar is a paradise. About 100 species of snakes are present in Madagascar, but there is no "true" venomous snake (although many of them are rear-fanged snakes with mild venoms). I have visited a tropical deciduous dry forest, Ampijoroa, in northwestern Madagascar for long-term ecological research, and we have recorded 20 species of snakes. There I can dive without hesitation toward a snake whenever it comes into my sight. However, I soon learned that something different is going on in that environment. If I dove toward the snake immediately when I saw it, I often failed to capture the snake. Because the body temperature of diurnally active snakes is very high in Ampijoroa, they can move quickly. Thus, if I dash toward a snake from a distance, the snake can evade my hand and escape into the bush before I reach it. Strangely, however, if I do not chase a snake, the snake does not escape into nearby bush. Instead, it stops fleeing after a short distance and stays still in an alert posture. If I remain motionless, the snake soon resumes its routine exploring movement. Therefore, I learned the following hunting method. When I find a snake fleeing, I should first freeze and wait until it stops moving. Then, after I visually locate the exact position of the snake, I carefully and very slowly approach the snake until I reach the range within which I can catch it by a short jump. After I started using this technique, the success rate of my hunting became very high.

Why do snakes in Ampijoroa adopt such a fleeing style? I speculate that it is attributable to the high density of animals in this forest. The Ampijoroa forest is not so large, having about 200 km^2 in area. Nonetheless, this forest provides home to many species of terrestrial vertebrates. We have confirmed 20 species of mammals, 86 species of birds, 55 species of reptiles, and 11 species of frogs—and some of these animals are living in quite high densities. Therefore, a given vertebrate would often encounter another animal, but in most cases the opponent would be a harmless coinhabitant that does not have to be avoided. Under this circumstance, if an animal flees a long distance to hide every time it encounters another animal, it might waste time and energy for unnecessary flight. Thus, natural selection might have favored individuals that make only a short run and determine whether the situation is really a risky one rather than individuals that run far away to a safe area whenever they meet uncertain danger.

There is no rule without exceptions. The exceptions are two relatively small, semifossorial species of Madagascar smooth snakes (*Liophidium*). They are usually crepuscular and are often found in leaf litter at the edge of small woods. If

A Japanese Striped Snake in a flooded rice paddy. To catch such a snake, we must carefully sneak up and quickly stretch out our hands while standing on the foot path because we are not allowed to step into the paddy water. Photo by Akira Mori.

I use the above freeze-and-sneak hunting technique against these species, I will lose them. These snakes almost always quickly disappear after a short burst of movement; they likely slide under the leaf-litter layer. For these snakes, I use the standard method: dive toward the snake as soon as I find it.

These observations in Ampijoroa are based on my personal experience and must be tested to be scientific evidence. Nonetheless, I am pretty sure that snakes have many different styles of fleeing, which would have been uniquely evolved in different species or under different environmental conditions. This idea would have never occurred to me if I did not go to the field and see wild snakes. Even in a simple behavioral act, such as fleeing, we find unexpected diversity. I am confident that there is more amazing, fantastic diversity of snake behavior that you can discover only when you go to the field and meet wild, active, healthy snakes. Now I can answer that this is one of the reasons why I am attracted to snakes.

About the Author

Akira Mori obtained his PhD at the Graduate School of Science, Kyoto University, in 1993. At present he is a professor of the Laboratory of Ethology, Kyoto University. He has been working on ecological and behavioral studies of Japanese snakes for more than 35 years. He has also led an international collaboration on a long-term ecological study of a terrestrial vertebrate community in a dry forest of Madagascar. His current major research topic is the evolution of a unique defensive system of Asian natricine snakes, which he is pursuing as another long-term international collaboration study with researchers from the United States and several countries in Asia.

21

IMMERSION

Erin Muths

It is late summer, the clouds over the Never Summer Range in north-central Colorado look like snow rather than a summer thunderstorm. I wonder if the metamorphs will survive. I can remember standing on the edge of that pond in the midst of a dark night in early May, throwing snowballs at the stump rising out of the slushy ice in the middle of the pond. Only meters off the road, the snow was thigh deep. No frogs were calling. The stars whined above in their winter way despite the calendar. The absence of amphibian sound and the twinkle of crystal remnants floating out of the blackness of the sky denied a promise of spring. I wrote that 26 years ago. That season gave me the pleasurable thrill of watching firsthand, and for the first time, what has been happening every year since the dawn of amphibians, about 300 million years ago.

The field work hasn't changed much, and what I wrote then is remarkably accurate (except that now I wear chest waders): *In June the water level lapped at the top of my hip waders. I had to watch my step as I grabbed calling males and watchful females. The delicate, unseeable, and unbearably loud chorus frogs frustrated me, and the metallic brown, triangular wood frogs enchanted me with their beauty and insistent call. They were mating, and it felt vaguely wrong to be watching.* Now, a quarter of a century later, I can better appreciate the profound wonder of the work but also recognize (and wonder at) the irrational fear that still sets my heart pounding when I take the bags of marked frogs back to their pond—really Erin? Two hours wading around alone in a pond in the dark, but you're afraid to walk 50 m back to the pond? Crazy. There are many trips where I'm alone, but on other trips there are people to help, new to the world of frogs.

Their enthusiasm reminds me of 1995 and the razor-edged excitement of stalking frogs that are so focused on sex that they don't notice my invasion of their tiny black ocean. The people all do their best, even when they are skunked. But some comments, voiced on frigid, early morning processing sessions, make me realize that not everyone shares a keenness for frogs. The most telling, from a criminal defense attorney and also my childhood friend, was "I would shoot myself in the head if this was my job."

It's late August now and the pond is gone, all the 200% above normal snowpack in the mountains has run off down the Cache la Poudre River. At the pond, the muddy floor sucks at my waders. Large invertebrates, a bizarre cross between shrimp and earwigs, wriggle amongst the grasses. And the froglets, they scurry, as only a tetrapod with a stump of a tail can, in and out of their moist, sedgey forest. Their numbers seem few, compared to the wall of sound produced by their parents in May, but they hide well, cloaking their bronze forms in leaf litter, dropping their lids demurely over riveting gold eyes. This population of chorus frogs (*Pseudacris maculata*) persists, this species apparently not poised on the brink of disaster. The wonder for me remains, but with it, a tiredness and a pinprick perception of futility in my life as a field biologist. I struggle with this feeling. But at the same time, I chide myself that these are the sentiments of the old, or those who can't immerse themselves in the watery glory of reflected stars and shouting frogs. I will immerse myself in their ritualized world again this spring, and I know that they have changed me in subtle ways that I may not even be aware of. This knowledge, that the natural world and my profession are intertwined, sustains me and provides a glimmer of optimism for the future.

About the Author

Erin Muths is a herpetologist at the US Geological Survey and a principal investigator for the Amphibian Research and Monitoring Initiative. She has studied demography, disease, and conservation in high-elevation amphibians for over 25 years. She has published more than 130 peer-reviewed papers and has served The Herpetologists' League and Society for the Study of Amphibians and Reptiles in various capacities including five years as coeditor for the *Journal of Herpetology*. Dr. Muths has mentored graduate students via formal university affiliations and informal interactions with technicians and junior colleagues. She received Southwest Partners in Amphibian and Reptile Conservation's Charlie Painter Award for mentoring in 2019. Erin's philosophy is that building capacity moves conservation forward and mentoring contributes to that.

22

HERPETOLOGY MOMENTS

Patricia A. Burrowes

Good teachers can make a difference and influence important career decisions in your life. I had a wonderful professor of herpetology in college (Marylyn Bachman) and a teaching assistant (Mike Lannoo) who made amphibians and reptiles so interesting that it was impossible for me not to fall in love with these animals for the rest of my life! I was raised in Cali, Colombia, where my parents had a home in mountains surrounded by cloud forest. My appreciation for nature, and my inclination to become a field biologist, probably came from my childhood excursions to the forest and early encounters with colorful snakes and frogs that frightened my friends but fascinated me with their body shapes and intriguing behaviors. Of the six kids in my family, I was the youngest and the only one who chose an unusual career.

Being interested in amphibians and reptiles and working in the forests at night to study frogs was not very typical for a woman at the time, and as a young herpetologist I faced several challenges. First, for my master's fieldwork I had to convince my parents to let me venture into the mountains of southern Colombia, where armed groups had started to rise. Also, the director of the La Planada Reserve where I did my work had no faith in me and made my life harder by banishing me to a building being built for future science visitors, claiming that I was the first. The walk from the "scientific center" to the reserve headquarters where meals were served was about 1 km, and at the time the building did not have electricity or running water. However, to his surprise, neither isolation, solitude, nor the constant pouring rain discouraged me from conducting my work. Every night I came back excited with my new discoveries of cool frogs that had unique reproductive strategies and of new "morpho species" of a very diverse group of arboreal frogs (genus *Pristimantis*) that were likely new to science.

One night, when returning late from monitoring amphibians in a transect, I heard a loud "BOP" very close to me. Bill Duellman, my major professor, had told me that there was a frog in northern Ecuador that lived in the high canopy and made a loud call like this. I had also been told by the native people in the reserve that this sound belonged to the cry of the spirit of a little girl that once got lost in the forest. I kept walking in the direction of the call, and then, right at my eye level, sitting in a tree branch and looking at me, I saw the most beautiful frog I had ever seen! It was a male Dentate Marsupial Frog (*Gastrotheca guentheri*). Female marsupial frogs have a dorsal pouch in which they brood fertilized eggs. Adult *G. guentheri* have a mustard-yellow coloration with longitudinal black markings and large eyes with horns on the eyelids. About a month later, I found a female very close to the ground; her dorsal pouch was full of eggs. I kept her in my cottage during the night and woke up to the clatter of this bagged frog jumping all over the floor. Every time she jumped a small emerald-green froglet with tiny horns on the eyelids emerged from her back. It is difficult to describe the feelings I experienced at the time, but imagine if you can, a mixture of the tenderness that comes with motherhood and the excitement of discovery—definitely one of the most thrilling moments of my life as a herpetologist!

But doing fieldwork in herpetology brings all kinds of moments. There were scary moments, like the time when I was almost bitten by an Eyelash Viper (*Bothriechis schlegelii*) in Costa Rica. I had the luck to share my Organization for Tropical Studies course with several herpetologists. We were all young and energetic and used all our free time to go herping. One day while measuring the diameter of a tree trunk, one of my team partners, Joe Slowinski, very calmly said to me: "Pacha, drop the meter and step back slowly." I did not question him. As I did as I was told, I saw a perfectly camouflaged Eyelash Viper about 20 cm away from where my hand had been. It is ironic and very sad to think that while Joe saved my life that time in Costa Rica, he died from a snake bite years later in Asia.

There were also "itchy" moments. Chigger bites made our lives miserable in Cuzco Amazónico (Madre de Dios, Peru), because our campsite was in a cleared area where pasture grew tall and provided ideal habitat for chiggers during the rainy season. Relief came when a colleague, David Hillis, joined us for a few weeks and brought a generous quantity of sulfur powder. It turns out that sulfur discourages chiggers from burying in your skin, so by mixing it with Johnson's talc and applying it in our socks and clothing, we were able to go from frantically scratching to stinking like rotten eggs all the day—totally worth it! David also brought Kit-Kats for me. I love Kit-Kats and hate beans (all colors), which constituted the core of our daily meals in the Amazon. So while they lasted, I had the luxury of complementing my limited diet of rice and papaya with loads of sugary calories!

I also experienced several near-death moments in the field. One I will never forget was in Colombia when I fell through a hole in the forest and into a stream in a steep canyon from which I was not able to climb out until daybreak—exhausted and suffering from hypothermia. However, what I remember most in my life as a field biologist are the moments of camaraderie that resulted from sharing exciting moments of discovery, despite adversities, with the many colleagues and students that became my friends.

I have been married twice, and both of my husbands have been herpetologists. My mother used to say that it was a great idea to marry someone from your same profession because you would never run out of conversation. Well, that is certainly true for Ignacio De la Riva and me, but I would add that it also helps having a spouse who understands your needs to work at night and often in remote, faraway places. Ignacio and I started working together in Bolivia to try to understand the drastic decline of amphibian species, many of which Ignacio had discovered and described in the late 1980s and early 1990s, when they were still abundant. We traveled all over the country to sample extant species of frogs for the pathogenic chytrid fungus, *Batrachochytrium dendrobatidis*, responsible for the decline of many amphibians worldwide. We also ran field experiments to test the vulnerability of high Andean species to climate warming. One of the highlights of these trips was seeing how Ignacio used Google Earth to predict the localities in the inter-Andean valleys where undescribed species of small, terrestrial direct-developing frogs from the genus *Microkayla* might occur. It was like magic to me, although, in reality, it was the result of careful study of this group of frogs and their pattern of speciation in the high Andes.

In 2016, we walked the Takesi route, a pre-Colombian trading path used by the Incas that extends from the high *puna* in La Paz to the cloud forests of Los Yungas. Ignacio had identified a specific location in the route where a new species of *Microkayla* might occur. We had booked a guide and a porter at an ecotourist venue at La Paz. The agency turned out to be incredibly disorganized and oblivious to what people might need when camping at high altitudes in the Andes. After many years of working in Bolivia, Ignacio knew better than to trust the agency operator when he said there was no need for us to bring sleeping bags, extra warm clothing, or snacks because they would take care of everything. If we had not brought our own share of all those things, we would have undoubtedly starved and/or frozen to death. Perhaps it was just coincidence, or perhaps there is a "God of Science," but on the first night we had to camp earlier than planned because it got dark and our guide did not want to continue the expedition. We were close but not at the exact GPS coordinates where Ignacio suspected the occurrence of a new species. We started to set up camp, and as dusk turned into night, we heard frogs calling. Ignacio knew it was one he had never heard before and most likely of an unknown *Microkayla* species. Excited, we looked around until we found the first frog under a rock, and then the second, and then a whole series including all age classes and both sexes, enough material for a thorough description of the new species! It was not the only time this happened to us in the Andes, and every time it was an incredible joy!

Working with students in the same place for many years gives a field biologist a sense of understanding of an ecological system that is beyond scientific data. Rafael Joglar and I began to monitor a highland population of the common Puerto Rican Coqui (*Eleutherodactylus coqui*), and the work has continued over 35 years. I have directed numerous graduate and undergraduate student projects, many of which led to theses and publications on the biology of these frogs. I have spent innumerable nights in the forest studying their behavior, movements, advertise-ment calls, mating sites, microbiome, and disease prevalence. And now, I feel that my hunches are educated guesses and that the knowledge I have acquired about this system allows me to keep formulating interesting hypotheses, without wasting time on unrealistic predictions. This feeling gives me a remarkable sense

Patricia Burrowes with Ignacio De la Riva, having just discovered a new species of frog (*Microkayla*) in the Andes of Bolivia. Photo courtesy of Ignacio De la Riva.

of accomplishment that comes only from studying animals in the field. Perhaps this is why I am reluctant to share my raw data when I publish the results of my work. Those brilliant scientists who have never gone to the field, but sit behind a computer and make impressive analyses of metadata collected by field biologists, will hardly have the criteria to reject results that make absolutely no sense in light of the species taxonomy, geographic location, season of the year, or interaction with other biotic factors intrinsic to the site where the data were collected. This is why, as much as I welcome technology and use it to facilitate my work, I feel that what a biologist learns by being in the field can never be replaced by automated remote sensors.

It is an interesting coincidence that I started to write this essay in La Planada, the place where I truly became a field biologist. After I finished my master's work in 1986, no other comprehensive study of the herpetofauna took place because the presence of armed groups in the area made it very dangerous to work there. The reserve was eventually shut down in 2007 after the director was killed on site by the FARC (Fuerzas Armadas Revolucionarias de Colombia). After the peace treaty in Colombia was signed in 2016, I saw the opportunity to fulfill my dream of returning to La Planada to re-do my master´s thesis. I wanted to see what had happened to the diversity of amphibians that I had described and quantified more than 30 years ago. Would the exacerbation of global climate change and/ or the arrival of emergent infectious diseases like amphibian chytridiomycosis have taken a toll? It took no effort to get Ignacio excited about this, and after considerable work we got the funding to return to La Planada. We brought with us a young graduate student, Claudia Lansac, who has agreed "to be me" and replicate my work from 1986. We have found that the cloud forest is intact and that the Awa native people, who are now in charge of the reserve, are committed to preserving the area and promoting scientific research. However, many species of frogs have disappeared (among them, my beloved *Gastrotheca guentheri*), and many of those that are extant are rare compared with 1986. La Planada is another example of biodiversity loss in the sixth mass extinction that we are so unfortunately witnessing today. Let us hope that science may lead us to identifying the culprits of these declines in the wild and of ways to reverse their action, so that among many other reasons, young herpetologists like Claudia can have their own herpetological moments.

About the Author

Patricia A. Burrowes is a professor in the Department of Biology at the University of Puerto Rico in San Juan. She received her PhD at the University of

Kansas under the mentorship of W. E. Duellman. Her early career research interests included community ecology, reproductive behavior, and population genetics of tropical amphibians. Since the eminent decline of amphibian populations worldwide, Patricia has dedicated her efforts to the study of the factors involved in this crisis, particularly the ecology of chytridiomycosis under enzootic conditions and the response of tropical amphibians to climate warming.

23

CRYING IN THE RAIN, IN THE MIDDLE OF THE WORLD

Luis A. Coloma

The 24th of February 2022, about 1800 hours. The dusk announces a dark night, and a soft drizzle embraces every living being. I am outdoors, in a restored montane forest of the Los Chillos valley in Quito, Ecuador. A chorus of *Cuico* frogs (*Pristimantis unistrigatus*) have begun to chant, and the "wrack tac tac" of a couple of male *Gualataco* Marsupial Frogs (*Gastrotheca riobambae*) heralds the coming rain. After walking a little ways from the forest, I find myself next to a fast-flowing stream, where an amplectant pair of a harlequin frog, the famous *Jambato* (*Atelopus ignescens*), walks at its shore—the couple is attempting to find the perfect place to deposit a clutch of eggs underwater. Nearly 10 m from the stream, I hear the calls of three other species of frogs once abundant in Quito (now nearly extinct): a Guard Marsupial Frog (*Gastrotheca pseustes*), Quito Rocket Frog (*Hyloxalus jacobuspetersi*), and Buckley's Glass Frog (*Centrolene buckleyi*). This walk is part of my routine and enviable life. I am lucky to be in the middle of the world, in a privileged equatorial hot spot of the planet Earth, the most "sapodiverse" country. Ecuadorians know that our country is a dream place for scientists, ecotourists, and frog lovers.

This magical scenario triggers memories of my first childhood encounters with frogs, more than half a century ago. The frogs I referred to above represent once common genera of a typical amphibian montane community in the Andes. In fact, at the end of the 1960s these genera and species were so abundant that for curious children it was irresistible not to take them home. Gualataco tadpoles were the first guests of our homemade aquariums, from which we discovered frog metamorphosis. The tails of these tadpoles had long been a source of medicine to cure the "red of the eye" (i.e., conjunctivitis). Edward Whymper, an Englishman and the world's father of mountaineering who climbed mountains in Ecuador in 1880, observed and collected the Gualataco and reported on its abundance: "In the vicinity of the town of Machachi we saw thousands, and must have heard hundreds of thousands of frogs . . . and in the evening their music was so loud as almost to interfere with hearing when walking out" (Boulenger 1882).

My childhood and early youth were full of discoveries, joy, and adventures with frogs in the Ecuadorian Andes. For example, and with some shame I must confess, I discovered, at about eight years old, the damaging side of the greenhouse effect when I collected dozens of glass frogs and placed them in closed plastic bags, fully exposed to the sun—and they died within a couple of minutes. What a fatal destiny in the city of Guaranda, the little town where I was born, located at the foothills of the Chimborazo volcano, the mountain where Humboldt, the father of ecology and biogeography, gathered data and received inspiration for his revolutionary scientific ideas.

Guaranda and the Andes have had a profound impact on my life as an amphibian biologist, taxonomist, and conservationist. As a kid, I discovered my first species of frog, a cousin of the black Jambato. One day, I was playing at the river margin and to my surprise I saw a bright red-orange frog (*Puca Sapo*) walk nearby. Neither its color, nor walking locomotion, nor day activity had been included in the only scholarly books I had at hand: *Escolar Ecuatoriano* and an encyclopedia depicting animals (frogs and toads were supposed to be green or brown, jumpers, and nocturnal). At first, I was disappointed to find that nothing I read or saw in the books seemed to fit the morphology or natural history features I observed. However, I was soon thrilled by the possibility of having found something new. About three decades passed before, in 2002, I published a paper describing this species (*Atelopus guanujo*) as new to science. Needless to say, it was a superb journey after studying general biology and later specializing in the systematics and ecology of amphibians at the prestigious University of Kansas and being mentored by two of the most dedicated and prolific amphibian biologists, William E. Duellman and Linda Trueb. Duellman and his students have been the taxonomists who contributed most to describing the astonishing number of 655 species of frogs from Ecuador, a country in which about 12 new species are being described every year—and no less than 200 remain yet undescribed.

The discoveries of species are still a work in progress, and more taxonomists and conservation biologists are urgently needed, given that the sixth mass extinction is occurring faster than expected and is converting some of us into forensic taxonomists. Sadly, the description of the Puca Sapo came nearly too late: in 2002 this species was presumably extinct and the last individuals were seen in 1988. Its disappearance was part of sudden declines, even in pristine and well-protected areas, that occurred especially by the end of the 1980s. Why did this unexpected disaster occur? Climate change and pathogens are to blame. This mass extinction is our fault. We, humans, became a population bomb, a term coined by the famous Stanford scientist, Paul R. Ehrlich.

Throughout my life, I have been a witness of frog abundance—but also of the declines and possible extinctions. Jambatos were so abundant at some sites that it was difficult not to step on them; they would walk during rains in the roadsides, even in disturbed habitats. *Jambato* is a Quechua name that means to cry in the rain. Given their profusion, they were the perfect animal on which to conduct experimental studies. In 1979, I conducted my first serious study (for a high school graduate thesis) of its thermal optimum and ability to resist low and high temperatures. I froze them in the refrigerator to see if they would survive. They did. I also put them directly in boiling water. Imagine the outcome! But, I also did interesting ethological observations and took my first photograph of a frog. I wrote in my high school thesis, "This frog seems to be very aggressive. . . . In the photo, it is observed that five of these are attacking one of the same species, but the latter differs from them in the color of the belly." I naively thought that the smaller ones were trying to kill the larger frog. Although I was so happy to have photographically documented this unique behavior, I realized later that they were trying to mate; the frog "being attacked" was a female and the five others were males. I would not even imagine that I would end up doing my doctoral dissertation studying the morphology and evolution of *Atelopus* frogs and later working, desperately, for their conservation.

In 1865, similar behavioral observations were reported by the well-known Spaniard amphibian biologist Marcos Jiménez de la Espada. Espada was the first to publish ample osteological data and natural history observations of amphibians occurring in Ecuador. It has been highly motivating to read his meticulous writings. For example, Espada (1875) gives an account (my English translation) of some aspects of the natural history of the once abundant *Atelopus ignescens*:

> We watched them by the thousands near swampy terrains during the months of November, December, and January in the grassy and humid meadows, near the streams, ponds, or lagoons. On the shores of the mentioned La Mica, in the Antisana, beginning in 1865, I surprised them at the time of their love and when the males look for the females to help them spawn or fertilize their eggs. They chased them through the swamps near water with activity and insistence, and so blind that, struggling to get them, when they reached them, they rolled in balls, scrambled with each other. I confess with regret that I then had little objection to what was offered me to more careful observation, contenting myself

with collecting and storing; for now, I have to reduce to mere conjectures what I could have recorded as a true fact. . . . We still have six couples in our collection, tenaciously embracing and just as I surprised them on the act. . . . Among the loose females there are several that still have very deep traces of the male embrace.

What Espada and I observed is also treated in a paper published in January 2022 in the journal *Animal Behaviour*. The young Colombian biologist Alberto Rueda and his colleagues detail similar observations known as "mating balls" as part of mate guarding behavior in harlequin frogs, *Atelopus*. Males may remain up to three months strongly embracing the female, whereas unmated males try to remove amplectant males from a female's dorsum or several males engage in physical combat to amplect a single female.

The end of the 1980s and especially the 1990s were a time of tragedy for frogs in Ecuador and in many parts of the planet. Many species became so rare that it seemed the end of the world for amphibians. I had in my hands, and photographed some, the last individuals of more than a dozen species that have not been seen since that time. That is the case for many species of *Atelopus*—or of an entire genus, such as the aquatic frogs *Telmatobius,* unseen in Ecuador since 1994. During the 1980s, I witnessed the collapse of the Jambato populations as well. In March 1983, I photographed what would be the last individual recorded in Quito, in the Chillogallo neighborhood, now fully urbanized. In May 1985, my parents and I observed thousands of road-killed Jambatos near Chimborazo; they were smashed while moving toward their breeding sites. In November 1987, Giovanni Onore and I collected some dead (for unknown reasons) animals 20 km east of Latacunga. However, I did not worry much at that time; they were so abundant that no one would think that they would disappear completely in 1988.

The surprising and sudden declines of frogs were a powerful reason why I had to move from taxonomy to conservation biology. In the 1990s, I began to build a Noah's Ark for frogs, which in a couple of decades has evolved into the *Arca de los Sapos* program, now being carried out at Centro Jambatu for Amphibian Research and Conservation. Remember the place (and frogs) I mentioned at the beginning of this essay? Yes, it is Centro Jambatu. Let's connect a few of those dots. At the center, the Cuicos and Gualatacos live freely in the garden. The first of these species has always been common, and the second is part of a reintroduction program. In contrast, the Jambatos are in an outdoor enclosure where we conduct assessment trials before attempting reintroductions. The calls of the three other species are sounds played in a touch-sensitive screen, where their photos are depicted as part of exhibits, in an effort to educate people about the familiar frogs that used to live in Quito.

An amplectant pair of the Jambato (*Atelopus ignescens*) from Chimborazo volcano, Provincia Tungurahua. The last individuals seen in Chimborazo in July 1986. Photo by Luis A. Coloma.

Returning to the Jambato, its story has a hopeful outcome. A schoolboy found a relictual population of Jambatos in 2016, in a remote location in Provincia Cotopaxi; a year later we managed to breed them under lab conditions, and now we are developing assisted reproductive technologies to store sperm and breed them to keep genetically viable populations. This is a unique opportunity to revive a beloved and emblematic species. Meanwhile, a thought for the future is recurrent in my dreams: Our center garden having six species of frogs living freely and announcing the coming of the rains, and millions of Jambatos wandering again in all the green areas of Quito, Latacunga, Ambato, paramos, and inter-Andean valleys, so we will have to walk carefully not to step on them.

References

Boulenger, G.A. 1882. Account of the reptiles and batrachians collected by Mr. Edward Whymper in Ecuador in 1879–80. *Annals and Magazine of Natural History*, Series 5, 9:457–467.

Jiménez de la Espada, M. 1875. Vertebrados del viaje al Pacífico verificado de 1862 a 1865 por una comisión de naturalistas enviada por el Gobierno Español. Batracios, Madrid: A. Miguel Ginesta.

About the Author

Luis A. Coloma received his PhD in systematics and ecology from the University of Kansas. He was a researcher and professor at Pontificia Universidad Católica of Ecuador. He is currently the director of the Centro Jambatu for Amphibian Research and Conservation (Jambatu Foundation). He leads the conservation program *Arca de los Sapos* to protect the Ecuadorian endangered amphibians. In 2007 he was the first recipient of the Sabin Award for Conservation of Amphibians, given by the World Conservation Union and Conservation International. In 2008 he was awarded the Saint Louis Conservation Award in recognition of his extraordinary lifelong dedication to the conservation of Ecuadorian biodiversity.

24

FROGS IN THE CLEAR-CUT

Julia E. Earl

I was crouched under some vines with my knees pressing into the soil. I carefully turned over clumps of dead leaves one after the other, feeling them crunch between my fingers. I could smell the leaf dust. My long brown hair fell in strands in front of my face and sweat rolled down my back. It was a warm September day, and I was searching for frogs. I peered through the vegetation looking for movement. Not seeing any, I rolled a small log and there it was! Something small and brown suddenly leaped away from the log. It was in one location and then suddenly somewhere else. The 3-cm-long juvenile Wood Frog (*Lithobates sylvaticus*) had such long legs that it could cover long distances (for its size) with a single spring. The Wood Frog's skin was metallic brown, almost bronze, and it had a pointed nose and a dark brown mask around the eyes. Its very long legs had subtle, splotchy stripes. The frog escaped my grasp rather quickly. Then it disappeared, blending in with the leaves covering the ground. I had to chase it down.

My doctoral advisor, Ray Semlitsch, had talked me into studying Wood Frogs. I had resisted because it seemed like everyone studied Wood Frogs and there were already loads of data on them, partly because they are the most widely distributed species of amphibian in North America. I had wanted to study something unusual or a species that no one had bothered to study because I've always been intrigued by things that are different and unknown. This might be because I've always felt a bit different. I am generally quiet and drawn to the trees. This made me stick out as a child but are rather good attributes for an ecologist.

Despite their widespread distribution and abundance in some areas, Wood Frogs are a species of conservation concern in Missouri, where I was working, because this population was at the very southern edge of the species' geographic range. By this time, I had also come to adore Wood Frogs. There was something mysterious about them. Adults appeared early in the year to mate and lay eggs when it was cold enough that I would wear a striped wool hat and bring a thermos of hot ginger tea to keep myself from shivering. It was strange to think the peak time for an ectotherm was when it was just warm enough for the pond's ice to thaw. Some other graduate students and I would go count the egg masses in each pond and sometimes find over 100 in a tiny pond where the surface of the pond looked lumpy with all that frog jelly. After mating, adult Wood Frogs vanished into the forest for the rest of the year. I saw only one adult outside the breeding season in the four years working there, and it was the weirdest thing to see it hopping through the forest in the middle of a hot day.

While searching for juvenile Wood Frogs, I was in a 3 × 3 m terrestrial enclosure made from a fine steel mesh, extending a few centimeters down into the soil and coming up to my waist. I had 24 of these enclosures stocked with newly metamorphosed Wood Frogs in June and July. The enclosures allowed me to keep track of these juveniles while studying their survival and growth. If you release a juvenile frog into the forest, even with traps, you will likely never see it again no matter how much effort you expend. By confining the frogs, I was able to find them again to collect data. I was using these enclosures to study the effects of different forestry practices on juvenile frogs; the enclosures were placed within a larger experiment on forestry practices. Circular areas were defined around amphibian breeding ponds, and each circle was divided into four pieces like a pie, with a different forestry practice in each piece of the pie. The four forestry practices included uncut, mature forest; clear-cut with logs left on the ground; clear-cut with no logs; and forest that had about half of the trees cut to allow more light, like a savannah. The two clear-cuts were always opposite each other, so if you looked down from an airplane, it looked like there were bow ties cut out of the forest. With enclosures in areas with these different forestry practices, I would be able to estimate whether the juvenile frogs survived better or grew faster in areas with certain forestry practices. I also counted the number of logs and measured temperature, soil moisture, and vegetation cover in each of the enclosures to better understand what was most important to juvenile frog survival and growth. If these variables were more important than the forestry practice, I might be able to make recommendations to foresters about how to harvest trees for a profit but still conserve amphibians on the landscape.

So, that September day I had to catch that frog to find out which one had survived and measure its length and weight. Given that I was in an enclosure and the frog couldn't escape, you would think that would be easy. However, I was also trying to avoid poison ivy to the face, the unrelenting thorns of blackberries and raspberries, and the never-ending radiating clusters of lone star ticks, each with a white circle on its back. I was a key victim of these ticks, frequently getting more than anyone else and earning the nicknames Tick Bait and Queen of the Ticks, neither of which I particularly liked. I had sent my volunteers to enclosures without poison ivy, since I didn't react to it (though I have since lost this fieldwork superpower, unfortunately). After scrambling around for a bit, I managed to grab my little frog.

Holding small frogs is tricky because you have to be firm enough that they can't wriggle their slippery skin out of your hand and escape but not so firm that you crush them. Wood Frogs are rather small and delicate as juveniles. I usually held them with my index finger and thumb at the "waist" so their long legs dangled down. I was sure this one was frightened to have a giant track it down, which made me feel a bit guilty, but I loved looking into their beautiful golden-flecked eyes and their hands held out like they wanted to shake mine in greeting. I inspected the frog's fingers and toes to find marks that individually identified the frog. I called out to my undergraduate research student, Paula, to write down the frog's ID number and the enclosure number on our data sheet, and then

Terrestrial enclosure in a Missouri forest that contains Wood Frog (*Lithobates sylvaticus*) juveniles. Photo by Julia E. Earl.

I placed the frog in a small plastic box with air holes in the lid and labeled it with the enclosure and the date. I would later take it back to the lab to measure.

In the other terrestrial enclosure in this clear-cut, we found a very large brown adult American Toad that was speckled with grey and rusty patches but not part of the study. When she saw him, Paula yelled "Sapito!" (meaning little toad) in her delightfully warm Peruvian accent and had the largest grin on her face. Paula loved toads because of their adorably grumpy demeanor. We gently let him go outside the enclosure and noted his presence on the data sheet. Then, we gathered up the gear, the box with the juvenile Wood Frog, and the other volunteers and moved on to the next set of enclosures.

As we walked toward the uncut forest, I looked back at Paula. She was carrying the clip board level and balancing the big toad on it, which was bouncing along in a seemingly contented way. I sighed and said, "Paula, put the toad back."

She looked at me disappointedly, "But he's so cute."

"Put him back. His home is back in the clear-cut."

She turned around begrudgingly. I went on to the next set of enclosures and got everything set up, so we could start the next set of searches. When Paula returned, we crouched down in the next enclosure, turning over every log, disturbing every pile of leaves, and investigating every hole in the soil.

We didn't find any Wood Frogs in that one. We wrote that down and went on to the next one all day long until we had searched all 24 enclosures and found 26 frogs out of the initial 400. We returned the next day to make sure we didn't miss any frogs. Of course, we did. It's very difficult to track down every individual in an area. In fact, ecologists expect that and actually estimate what's called the detection probability to account for it in statistical models. We found four more juvenile Wood Frogs the second day, along with a beautiful Earth Snake and a Leopard Frog. It turned out that we must have missed another three, because we returned the following May to look at survival and growth over the winter and found a few frogs that weren't caught back in September. Paula worked hard with me every step of the way, spreading joy and generally making the work more enjoyable. I baked chocolate chip cookies, which she loved, to help keep her happy during the long work days. She often placed the cookies on the dash of my car to make them all warm and gooey. I brought watermelon for everyone as well, which helped cool us down after working in the heat.

In the end, we found that juvenile Wood Frogs had low survival, which is fairly common for young frogs. Only 12% survived from metamorphosis in June until September. For the frogs that made it to September, they had a 39% chance of surviving from September until May. The one thing that really helped the little frogs survive was logs. Having logs on the ground helped protect the soil from drying and gave the frogs a protected area to shelter from heat and predators. Many insects also like logs, so the logs may have provided a buffet of frog food. The presence of logs was actually more important than whether the area had been clear cut or not. My study was done three years after the initial forest harvest, so the clear-cuts didn't look quite like you might imagine. By three years after the trees were cut, small trees, bushes, and brambles had grown up to my shoulders or higher. The presence of this vegetation shaded the ground enough that there was no difference in the temperature or soil moisture between clear-cuts and forest. It still got really hot and dry in summer in Missouri. My study showed that retaining some logs on the ground after harvesting timber could increase the survival of Wood Frogs. This may not be appealing to foresters because that might represent some loss of revenue, but they could choose damaged or crooked trees that are worth less as a way to help conserve Wood Frogs. To me, it's a small price to pay to see those charming frogs with the bandit mask, bronze skin, and golden-flecked eyes.

I haven't been to that site in Missouri in 10 years, and it was cut 18 years ago. I sometimes pull up the satellite images on my computer so I can see what it looks like. If I look carefully, I can still see the faintest of bow ties in the forest, where the trees are different heights from that clear-cutting. I wonder how the Wood Frogs are doing, and I miss my time in the forest with Paula and all my other friends who helped me catch frogs. I still catch frogs and more often listen to them sing, but there aren't Wood Frogs where I live, and it's not quite the same.

About the Author

Julia E. Earl is an assistant professor in the School of Biological Sciences at Louisiana Tech University. Julia and her lab of master's and undergraduate students study the connections between ponds and forest and the conservation and ecology of amphibians and aquatic insects. She teaches courses on environmental sustainability, conservation, and freshwater ecology at both the graduate and undergraduate level. Julia lives with her husband, Dylan Allen, and their two children, Lucy and Andrew, in the small town of Ruston, Louisiana, where she likes to cook, garden, and hike.

25

ONCE UPON A DIAMONDBACK

Learning Lessons about the Fragility
of Desert Life

Wolfgang Wüster

Life is both remarkably resilient and at the same time scarily fragile. It persists in its numerous forms in the most hostile environments, yet its constituent components are often extraordinarily sensitive to even minor disruption of their habitats. Individuals in many environments live life on the edge, precariously balancing the physiological limits that govern their raw survival with the equally strong drive to gather the resources needed to ensure reproduction to pass their genes on to the next generation. In extreme environments, this balancing act can be precarious enough for even very minor chance events to make the difference between life and death.

Intellectually, messages on the fragility of life are easy to understand. We are bombarded with messages on the delicate nature of the intricate ecosystems surrounding us and their vulnerability to disturbance. For those interested in reptiles and amphibians in nature, field herpetology Facebook groups are full of reminders that disturbing animals can have profound consequences for their survival and that of their populations. However, while these reminders are easily read, they are even more easily forgotten, without triggering deeper assimilation or reconsideration of one's own field methods and ethics. All too often, it takes the personal experience of witnessing this fragility and the consequences of seemingly minor events to develop a deeper consciousness of this vulnerability.

One of my moments of epiphany on these issues came from a chance encounter with a Western Diamondback Rattlesnake (*Crotalus atrox*) in Cochise County, Arizona, on a fine morning in September 2018. After turning northeast off Highway 80 in the San Simon Valley toward the majestic Chiricahua Mountains, illuminated by the early morning sun that was now behind me, I kept scanning the dirt road ahead for signs of life, especially snakes, while heading toward the mountains for a photographic project. As is common in this part of the world, I soon spotted an average-sized adult Western Diamondback Rattlesnake emerging from the roadside vegetation on the right-hand side of the road, seemingly from a rocky outcrop located a few meters from the road. As the snake slowly progressed across the road, without taking much notice of either the four-wheeled creature now stationary a few meters away or the delighted bipedal synapsid it had disgorged, I had a remarkable view of the snake in the foreground gliding across the road with the majesty of the Chiricahuas behind it. I seized the opportunity to lie flat on the road at a respectful distance and got some photos of the scene. Apart from a few brief stops and tongue flicks, the snake paid no heed to my presence and continued to cross the road until it disappeared into some bushes bordering fenced-off pasture to my left.

Western Diamondbacks are extremely abundant snakes in many parts of the southwestern United States, and chance encounters like this are almost daily occurrences. Nevertheless, I never tire of the majestic sight of one of these creatures crossing a road. Delighted with the impromptu encounter, I continued on my journey toward my intended photographic project, which went off without a hitch. Thus satisfied, I started my return drive back toward Highway 80, this time driving against the sun. As I passed the location of the diamondback sighting approximately two hours after the initial encounter, the backlighting brought out in spectacular fashion the tracks left by the snake earlier that morning. I stopped and took a few shots, showing the tracks' origin in the rocky outcrop in the background and its rectilinear progress across the dust and gravel of the road.

It was only while photographing these tracks that I noticed, somewhat to my surprise, that their author had emerged from the bushes a few meters away from me and was heading back onto the road, presumably toward the outcrop from where it had emerged earlier that morning. However, something was strikingly different: the snake now contained an enormous bulge, and was moving slowly and with obvious effort. "Waddling" is not among the commonly accepted forms of snake locomotion, yet it seems the most apt description for the gait displayed by rattlesnakes after eating very large meals—not unlike that of herpetologists after a late-night visit to a Denny's. That certainly applied to this particular diamondback.

As I watched the snake and its bulge, several thoughts shot through my mind. That snake had headed out to a clump of bushes with an empty stomach and was now attempting the return journey a mere 90 minutes later with a full stomach. Ninety minutes! Barely the time required to find, kill, and eat the kind of large meal this snake clearly contained. Some of my own shopping trips to the local

Outbound and inbound. Left: Western Diamondback as first seen, heading across the road. Right: the same snake, after aborting its attempt to cross back toward its rock outcrop after feeding. Note very large food bulge, which was absent on the outbound journey. Photos by Wolfgang Wüster.

supermarket, carefully planned to minimize the time spent in an environment that I consider an antechamber to hell, take almost as long. Did that snake really plan its gastronomic foray with this level of foresight and strategy? Can predation success really be this predictable? Did the snake "know" what it was doing and "plan" the timing of this outing? After I related the experience on social media, suggestions included that the snake was following a prey item it had previously struck and released. However, the photos of the snake tracks across the road revealed no evidence of a large prey mammal having preceded the snake on that journey. Similarly, in the absence of traffic, it seems unlikely that it was aiming for a roadkill, even though rattlesnakes are known to sometimes consume these. Yet, if this was really a "planned" foraging outing, that raises questions on the nature of ambush versus actively foraging snakes. What would I have given to have observed the full sequence of events that unfolded after my earlier morning encounter with that snake. . . . I found myself wishing that I had scrapped my photo project and followed that snake instead earlier that morning! But that's hindsight for you.

I was awoken from my musings by the fact that, unfortunately, my presence disturbed the rattlesnake and dissuaded it from continuing its journey: it turned around and headed back into the bushes from which it had just emerged rather than continuing toward its intended destination on the opposite side of the road, the rocks from where it had come this morning. Very slowly, because of the size of its bulge, it "waddled" back into the bushes.

At this time, midmorning, the sun was already high in the sky and temperatures were rising rapidly. The snake was clearly keen to get back to its rocks, but
I had gotten in the way. I felt the first pangs of conscience about the entirely
unintended impact of my presence on its plans for the day and immediately
decided to retreat to allow the snake to resume its journey toward its intended
destination. I got back in the car and parked several tens of meters away from
the snake and the morning's crossing point and waited to see if the snake would
try crossing the road again. Concern mounted that it might end up spending too
much time in the sun owing to the burden of its meal and risk overheating on
the shelterless road. However, after approximately 30 minutes, the snake had not
re-emerged from the bushes.

A quick look under the bush confirmed that the snake was still there, now
loosely coiled in the dappled shade of one of the bushes. I considered my options.
Leaving the snake alone, to make its own decisions, was one. The other was capturing it and returning it to its rock outcrop. The latter would have involved
significant additional stress, and the risk that the snake would regurgitate its large
meal, a significant loss of energy. Moreover, there was no way of knowing *where*
among the rocks that snake had started its journey or the thermal qualities of
the various available shelters in the outcrop. Capturing and returning the snake
without a clear destination seemed risky.

Leaving the snake in place also carried risks. However, after due consider-
ation, in terms of temperature, the situation seemed manageable. Daytime temperature maxima in the San Simon Valley at that time of year are usually in the
low–mid 30s (°C)—well within the envelope of survivable temperatures for a
Western Diamondback. Moreover, the snake had access to the dark shade of a
very dense bush right next to the one it was sheltering under at the time. Consequently, I felt confident that the snake would be able to sit out the heat of the
day under that bush and decided against capturing it and carrying it back to the
rock outcrop. But, of course, that was still a suboptimal solution: in the bushes,
the snake was at a considerably increased risk of exposure to predators, and
there was the possibility that the snake might become a victim of road traffic
during its next crossing attempt. With some anxiety, I left the snake to its own
devices under its bush.

Through the rest of the day, my thoughts returned frequently to that diamondback, with concern for its welfare. That concern continued through the
night, when, as so often, the imagination ran riot during moments of early morning wakefulness, conjuring up the worst possible scenarios. Since thoughts of that
diamondback and my impact on its day never fully left my mind, I sought closure
the next morning and returned to the scene. Words cannot describe my relief at
finding a second, new set of rattlesnake tracks across the road. This new track

headed from the bushes under which my waddling snake had been sheltering back into the rocky area by the side of the road. The space under the bushes was devoid of any sign of the snake as well as any signs of a struggle or regurgitation: the snake had made it back home!

So this chance encounter had a happy ending for everyone. And yet, the outcome could so easily have been different. Had the bush under which the snake sought refuge not been dense enough, or the day that little bit hotter, my inadvertent disturbance that stopped that snake from reaching its intended shelter could have been a death sentence. The same encounter in the lower-lying areas of the Sonoran Desert, where daytime temperatures often exceed the lethal maximum temperature of a rattlesnake, would have ended fatally if the snake had not found a sufficiently deep refuge that it could reach and fit into despite its meal. Equally, left in the open rather than the shelter of the rock outcrop, the diamondback, slowed down by the bulk of its prey, would have been easy prey for any passing predator. My simple presence and unintended disturbance of that snake, without any deliberate interference, much less actual contact, could easily have made the difference between its survival and its demise. Good fortune rather than design prevented my mere presence from spelling the end of that beautiful diamondback's existence.

The take-home message? Small disturbances can have big effects. Desert snakes need to balance the need to forage, the need to avoid dehydration, and especially the need to avoid overheating in the daytime. Small disturbances that seem trivial to the casual observer could tip a snake past the point of no return, by causing it to overheat directly or by preventing it from returning to safety. Those warnings offered by keen field herpers on Facebook and elsewhere of the impact of seemingly minor disturbances are right—small actions can have big consequences. So my own take-home message from the episode was to reflect more on the fragility of the life I so admire, and my impact upon it, however unintentional it may be. By sheer stroke of luck, I can remember this encounter as a lesson learned without harm to anyone and with a smile on my face. But I will never forget how easily it could have turned into a moment of shame that would have stayed with me for life.

About the Author

Born in Germany, **Wolfgang Wüster** is a professor in zoology (herpetology) at Bangor University, North Wales, United Kingdom. He obtained his PhD on the systematics of Asian cobras at the University of Aberdeen, Scotland, in 1990, and has since developed a broad research program on the ecology, evolution, and natural history of venomous snakes and the role of their venoms therein, as well as their conservation. These academic interests are combined with a passion for nature and wildlife photography. While he has had the privilege of carrying out fieldwork on all continents except Antarctica, he has developed a special affinity for the deserts of the southwestern United States and their buzzing residents.

26

SWAT TEAM TO THE RESCUE

Joan (Diemer) Berish

Conducting research on Gopher Tortoises (*Gopherus polyphemus*) in Florida for 33+ years provided me with a wealth of information about inherent impacts on this threatened species. Not unexpectedly in a state that averages around 6000–7000 new residents per week and many thousands of tourists, anthropogenic impacts are extremely common. Unfortunately, the list is long and the toll staggering, for example, development, agriculture, mining, thickly planted pine plantations, forestry machinery, human predation (although now illegal), vehicles, and dogs.

But Florida also has a number of natural environmental factors that can affect this burrowing turtle: predation (especially on eggs and juveniles), hurricanes, fires, floods, and a lesser known one, sinkholes. As I traveled around the state to assess regional and localized impacts on Gopher Tortoises, sinkholes really didn't cross my mind. Yet, it's an understatement to say that Florida is famous for its sinkholes, which have swallowed roads, pools, homes, and even people. It doesn't take much extrapolation to reason that wildlife, too, can run into problems with these deep depressions. Why Florida? Much of the Sunshine State is karst terrain, that is, largely underlain by limestone. Rainfall, especially acidic rain, can percolate down and dissolve the stone, creating caverns and underground spaces.

This prelude sets the stage for an unusual tortoise rescue that took place in August 2000 west of Gainesville, Florida. A local resident out walking her dog in an abandoned field stumbled on a sinkhole partially obscured by saw palmetto (*Serenoa repens*) and dogfennel (*Eupatorium capillifolium*). To her surprise and dismay, she spied a Gopher Tortoise at the bottom of the approximately

30-ft-deep × 20-ft-wide hole. Her first instinct was to get food to the tortoise, so she dumped lettuce and cabbage leaves into the abyss. She considered trying strawberries but thought they would attract too many ants. She also called a local wildlife rehabilitator who then called the Florida Fish and Wildlife Conservation Commission, and that is where I came into the story. I quickly coordinated with my colorful colleague, Ray, who just happened to have a tortoise research and conservation area not far from the site of the trapped tortoise. When Ray and the lady who discovered the stranded tortoise peered into the hole, the tortoise did not move. Even tossing small sticks down upon it did not elicit any movement. After observing this atypical situation for a while, they both deemed the unfortunate creature a "fossil," as in "way dead." There was really no way to know when this large tortoise tumbled into the sinkhole and how long it had been there.

Almost a week later, the lady was again walking her dog and decided to take a quick look, tossing more sticks into the sinkhole. This time, the tortoise moved! She immediately called the wildlife rehabilitator who then called me. I coordinated directly with the witness to this dilemma and arranged to meet her at the site. My field notes were succinct regarding the sinkhole: "deep and creepy." But the tortoise was indeed alive, and now I needed to figure out how to retrieve the poor animal quickly and safely from its predicament. I conferred with several local firemen from this small community, and they suggested that I contact our county fire and rescue guys who had some experience rappelling. Alas, all their rappellers were out West fighting wildfires. I made numerous phone calls and scrambled to try to find someone who might have rappelling expertise and could be available as soon as possible. Running out of options, I explained this challenging scenario to one of our wildlife officers, Captain Cook (really!), who had a brainstorm. He called the county sheriff's office because he knew that their SWAT (Special Weapons and Tactics) team had just returned from advanced rappelling school, where they had descended the Alltel Stadium in Jacksonville. Now this was a case of propitious timing. Captain Cook made the valid point that the SWAT team members could use more practice. Heretofore, the sheriff's department had rescued a few cats—and even some escaped cows—but this was a critter of a different color, so to speak.

We needed to shift this undertaking into high gear, so I rushed back from my ongoing tortoise field research two hours away and met the deputies at the sinkhole on a hot summer afternoon exactly one week after the tortoise had first been seen. In my field journal, I noted that this was a good team effort, if not somewhat of a three-ring circus. Our rescue crew consisted of the three SWAT team members and the sheriff's office spokesman, three local firefighters, the lady who discovered the tortoise and her friend, a newspaper reporter and his photographer, a

TV news cameraman, and myself. Initially, they wanted to try lowering an open-ended cage on a rope. "I got the door hitting him right on the butt," exclaimed one of the deputies. Alas, this tortoise was having no part of that cage that swung near its shell. It was time for Plan B. With the help of the local firefighters, one of the deputies rappelled into the deep sinkhole. He quickly recovered the tortoise, placed it in a burlap sack, and sent the rescued reptile upward bound on a rope. The hero deputy was then expeditiously hauled back to the surface: most folks don't want to spend much time in a creepy Florida sinkhole.

Now it was my turn to take the reins of this operation. As the state's tortoise research biologist, I would be responsible for assessing this animal's condition and finding it a more appropriate location. It was a good-sized adult male, who may have been out foraging, or looking for love, when he tumbled into something much deeper and perilous than his burrow. Late summer is the time when

Lucky the Gopher Tortoise was dehydrated, but still kicking, when he was rescued from a Florida sinkhole. Photo courtesy of Terri Tucker, taken of the original *Gainesville Sun* article photograph.

males can get a wee bit frantic trying to find available and willing females with which to mate. I knew that the tortoise was dehydrated, so I placed him in a shallow plastic tub and poured in an inch or two of water. This now fortunate tortoise, named Lucky by the deputies and lady who discovered him, took gulp after gulp of water and seemed to enjoy the soak. We watched in fascination as his eager drinking stretched the folds of skin on his neck. The spokesman for the sheriff's office officially declared that "Operation Tortoise 2000 is a success," and the local newspaper ran a story about the rescue, complete with photos of the SWAT team in action and, of course, of Lucky. The newspaper story noted: "When it was all over, Lucky just needed a good, long drink."

My buddy Ray was thrilled to hear about the rescue but also chagrined that he thought the poor tortoise had succumbed. I assured him that it must have been difficult to discern a tortoise's condition from 30 ft above. We were both thankful that Lucky had survived. I've always said that tenacity and perseverance are admirable tortoise traits. It also helped that tortoises are relatively well equipped to go without food or even water for some period of time; they are adapted for long stints underground. But this wasn't the safe haven of his burrow, and even the heartiest can take only so much.

I had Lucky checked over by a veterinarian who had experience working with tortoises. Lucky had a few abrasions from his ordeal but, once rehydrated, seemed to be no worse for wear. He was also tested for upper respiratory tract disease and was deemed negative. Although there was at least one burrow near the sinkhole, there was no way I was going to release him back in that field fraught with danger. Moreover, this field was rumored to be slated for eventual development. Fortunately, my colleague Ray welcomed Lucky into his vast and well-managed tortoise sanctuary. Gopher Tortoises are renowned for being long-lived, so Lucky may still be roaming through those sandhills characterized by towering longleaf pines (*Pinus palustris*) and brilliant green turkey oaks (*Quercus laevis*). And hopefully, he is avoiding deep holes unless he digs them or takes over another gopher's burrow. One might say it's a jungle out there in the Florida sandy lands: every tortoise for himself, unless of course you need the help of a SWAT team.

About the Author

Joan (Diemer) Berish began studying sandhill reptiles in 1978 after working in nuclear medicine research, veterinary science, and wildland firefighting. In 1981, she received her master's degree in wildlife biology from Auburn University; her thesis was on the distribution of the Eastern Indigo Snake in Georgia. From 1980 until her retirement in 2014, Joan was employed at the Florida Fish and Wildlife Conservation Commission's Wildlife Research Lab in Gainesville. Her primary research over three decades involved Gopher Tortoises. Her memoir, *Fire and Fauna: Tales of a Life Untamed*, was published by Texas A & M University Press in 2019.

27

MILITARY HERPETOLOGY

Robert Lovich

The year was 1999. My coworker Drew Stokes and I were finishing up a day of checking pitfall arrays scattered across Marine Corps Base Camp Pendleton as part of our work with Robert Fisher at San Diego State University. Our survey work included locations all over southern California. That afternoon, while walking to our last location, we stopped at the junked-out body of a VW beetle to flip it over and say hi to the Speckled Rattlesnake that was almost always underneath. There she was, and it was nice to see familiar wildlife still there as expected. We heard helicopters flying in pretty close but thought nothing of it. Across the base we would normally encounter low-flying helicopters, amphibious vehicles, Humvees, and the familiar sounds of various small arms, munitions, and occasional larger explosions in the distance. This was all part of the landscape for military testing and training. What was interesting to Drew and me was the pattern that was emerging from our data across southern California. Military installations had the most species and the highest densities of species being captured. Drew and I always looked forward to doing a round of trapping and surveying on an installation because we knew we would find lots of great stuff that few others could see.

That afternoon, as the helicopters landed nearby and we finished processing the last captured animals, we walked back to the truck enjoying the long shadows of another coastal California sunset. Then, all hell broke loose, and it sounded like a rolling thunderstorm erupted inside a volcano! The helicopters were bringing in an artillery squad to practice flying in, setting up, and opening fire on a target with 155-mm artillery. We heard the shells whizzing overhead as they flew from the canons. Since we were in hilly oak woodland with canopy overhead, we had no idea that we were only a few hundred meters from the landing zone and artillery that went live. After Drew and I checked our underwear, and after our ears finished ringing, we high-fived each other! It was exactly the kind of experience he and I relished while working on base. And then we safely got the heck out of there.

In 2001, I was hired as a wildlife biologist for Marine Corps Base Camp Pendleton and my career began as a biologist in the military. In 2006, I transitioned to working for the US Navy at what was Southwest Division at the time. Over the years, my work has included riding in helicopters to remote locations; using handgun-propelled flares to start backfires to contain wildfires; engaging the combat engineers to use their C4 explosives on trees to provide a training opportunity and thereby save money on exotic nonnative tree destruction/removal; wrangling Burmese Pythons in the Florida Keys where the US Navy is the largest landowner; driving out to the firing range to see a first-of-its-kind shoulder-mounted missile launch up close; surveying for and capturing feral horses and burros by herding them with helicopters in the air and cowboys on the ground; and numerous other worthy adventures. What is my job? I am a senior natural resource specialist for the US Navy and the national technical coordinator for the Department of Defense Partners in Amphibian and Reptile Conservation (DoD PARC) network.

Fast forward to 2022, and that year finds Chris Petersen and me engaged in drafting a scope of work for Western Pond Turtle (*Actinemys marmorata and A. pallida*) surveys on every military base on which they occur, developing best management practices documents for mission-sensitive species of herpetofauna, developing partnerships related to herpetofauna on military lands, updating and maintaining the DoD PARC herpetofaunal free photo website, maintaining the database of all amphibian and reptiles species on approximately 425 military installations, developing species profile videos for the internet, providing educational outreach, developing DoD PARC podcasts, analyzing passive acoustic frog and toad calls, and developing a coloring/comic book for amphibians and reptiles on military lands. This is a mouthful, and the result of many years of hard work growing our capacity within the DoD for amphibians and reptiles. We are also managing nationwide snake fungal disease surveys on nearly 100 military bases across the United States. These broad-scale planning, programming, and scoping efforts are a normal process for DoD PARC; we are tasked to secure the funding and support needed to sustain the military mission while at the same time striving to protect, manage, and study amphibians and reptiles on the 25 million acres of DoD lands in the United States. We also provide oversight for similar efforts conducted on military lands overseas.

How Did We Get Here?

Let's go back to 2006. I was being hired by the US Navy in San Diego, California, after working for the US Marine Corps. Staff and leadership at navy headquarters and the DoD in Washington, DC, were aware of a doctoral candidate named Rob Lovich as well as another recently graduated herpetologist who had started with the US Navy in Norfolk, Virginia, Chris Petersen. Both the navy and DoD were determined to initiate a military-wide network for amphibians and reptiles, primarily because military lands are home to the highest density of threatened and endangered species of any federal lands in the United States. Amphibians and reptiles follow that trend with tremendous densities and diversity, regardless of their status as protected or not. The US Navy and DoD recognized the benefit of getting all the military services (US Army, US Navy, US Air Force, and US Marine Corps) to work collaboratively on amphibian and reptile conservation and management, especially for military bases that shared the same protected species but were not necessarily working collaboratively to protect and recover those species.

Chris and I were encouraged individually to begin drafting a strategic plan for the entire DoD for the conservation and management of amphibians and reptiles. Chris and I had never met each other, but we enthusiastically started working on this enormous undertaking to bring together and harmonize the military services natural resource programs and to adopt the shared goals and multiagency coordination efforts of protecting, managing, and researching amphibians and reptiles across all military lands in the United States. By 2007, we were submitting funding proposals to DoD. By 2008, we received our first funding to create the DoD Strategic Plan for Amphibians and Reptiles, as well as to sample the continental United States in a single transect for amphibian *Batrachochytrium dendrobatidis,* or BD, which causes the disease chytridiomycosis. While creating and implementing the strategic plan was our priority, the scientific community, at the time, was grappling with the devastating effects of BD on amphibians. We felt it was imperative to assess the effects of BD on amphibian species on US military lands in order to protect both the species and the military mission. Our reasoning, then and now, is that amphibian population declines, or population declines for any species for that matter, can result in greater legal protections for those species. Those protections can lead to increased management costs and increased permit and consultation requirements with regulatory agencies, which can encumber the military mission. Thus, we recognized early in our careers the benefits of proactive conservation to avoid species declines and also the need for multistakeholder collaborations (across the range of the respective species) to help reverse those population declines.

Chris Petersen and I received our first year of funding in 2008 and embarked on our ambitious DoD PARC nationwide BD transects with Mike Lannoo, Priya Nanjappa, and Joe Mitchell. Joe did all the sampling, and the rest of the team took turns assisting him at various military bases over the course of the study. From the coast of Virginia to the coast of California—and at 13 other military bases in between—we sampled all amphibians encountered at three different intervals that first year. Those preliminary results were published under the title "Do Frogs Still Get their Kicks on Route 66." It was a great start that was well received and provided vital new scientific information. Additional studies followed, with transects covering the United States from north to south and a citizen science effort that included several dozen bases all across the country. Study findings revealed a general pattern: BD is endemic to the United States; it is not causing widespread declines of species in the United States as it has in other parts of the world; and BD is generally less common in hotter and drier regions of the United States.

When considering the DoD strategic plan, we looked first to the model of the national nongovernmental organization called Partners in Amphibian and Reptile Conservation (PARC). They already had considerable size and depth and had brought together hundreds of stakeholders for the shared goal of conserving amphibians and reptiles. Why reinvent the wheel, right? Chris and I worked collaboratively with Priya Nanjappa and Ernie Garcia; their guidance was particularly helpful with drafting the plan and formulating how to gain support for ultimately implementing the plan. Chris and I then led a series of strategic plan workshops for military leadership to review and discuss the plan and to gain their support and concurrence. Let me tell you, that was no easy chore. Implementing a bold new initiative for all of DoD is a task earned and not given. Myriad considerations need to be made, and frankly, people need to be convinced the effort is worth implementing for a better outcome.

Each of the military services already had their own natural resource programs and were spending enormous amounts of money protecting their respective amphibian and reptile species. They had valid concerns about funding yet another department-wide network specific to amphibians and reptiles. I remember one 2010 meeting in particular, when I was giving a strategic plan presentation in front of military leadership. I was extremely nervous beforehand, and a dear colleague pulled me aside backstage and said, "You need three things for a successful project of this kind: (1) Funding, which we have given to you and Chris. (2) A plan, which you have drafted. And (3) a hero. You're that hero, now go out and do your job." Her words immediately affected me, and I knew that whatever my own feelings, this was bigger than Rob Lovich or any other personality involved, and I had to go out and lead this effort.

Trust me when I say there was plenty of work for Chris and me to share, and plenty of times we each had to put on our proverbial superhero capes to maintain the support needed to build such a vast effort for an entire department of the US government. But in that moment, I stepped into my role and led that presentation. I stood the test, and we delivered the goods. Several more workshops, and enormous and complex rounds of tense negotiations, followed. Even though I grew up in Virginia, literally across the Potomac River from Washington, DC, developing our strategic plan was a "baptism by fire" in what they call 'Beltway politics' in our nation's capital and in what it takes to build something this expansive for the entire DoD!

Fast-forward to 2015, and we were able to finalize the DoD Strategic Plan for Amphibians and Reptiles. There were literally years of delays, intervals of nonconcurrence by some of the military services, lots of hand-wringing, vast amounts of time spent advocating/assuaging individuals, and endless reviews and re-reviews before it was finally complete. When we put it in front of John Conger, acting assistant secretary of defense in 2015, he signed the plan with no objections and the full support of the Office of the Secretary of Defense, Army, Navy, Air Force, and Marine Corps, and we were finally able to fully implement the plan. Our internal processes now changed completely with an authorized strategic plan to implement. New activities such as branding, creating a logo, developing discrete products, and undertaking projects to assist the military services with amphibian and reptile everything were underway. It took six years, but now we are fully legitimate. I have to take a moment to share my appreciation for the foresight, support, and hard work that the staff at the DoD have given DoD PARC all these years.

Disease

As mentioned, some of our formative work to address threats to amphibians and reptiles on military lands included several projects focused on BD. As Chris and I have managed DoD PARC through time, various wildlife diseases have continued to arise as threats, and we have been forced to tackle such things head-on. Neither Chris nor I went to school to become experts in wildlife epidemiology per se, but boy have we had to add that to our skill set. It's a terrible indicator of global biodiversity conditions, but BD, *Batrachochytrium salamandrivorans* (Bsal), snake fungal disease, turtle shell disease, among others, have all had devastating effects on herpetofauna and appear to be increasing. It isn't something I signed up for, nor is it a particularly appetizing topic of research, but continued research and abatement for mitigation of the deleterious effects from these pathogens is of paramount importance.

Fieldwork/Fun

Have you heard the saying, "Find a job you love and you'll never work a day in your life?" As I have told many of my colleagues over the years, I am thrilled each day by the fact that I have a job in my favorite profession (biologist–herpetologist), AND I get to be a patriot serving my government as a civil servant and employee of the US Navy. The larger military landscape where I conduct my work is an amazing place with sheltered and protected biodiversity, including amphibians and reptiles, as a direct result of having natural and secure lands for the training and testing mission. Chris and I, and our other military biologist colleagues, get to go all over the world managing, protecting, and studying the amazing species that reside on our lands.

Our inventories of amphibians and reptiles have covered the globe. Chris and I get to do a lot of the work, but our colleagues stationed at military bases within and outside the continental United States also do a lot of the heavy lifting. Whether conducting general inventories for species literally all over the United States, or traveling to places like Sicily, Japan, Kenya, Spain, Greece, Romania, and Mexico, we get to seek out and define the composition and status of herpetofauna wherever the US military needs it done. We even revised, with the help of many colleagues, the venomous snakebite treatment protocol for the DoD in the European and North African theater of operations and wrote a field guide to the venomous snakes of several countries in Central America, all at the request of the military services. You never know what we will find in our inbox, or where it is coming from, to serve the men and women of America's military services better. It has been especially rewarding as well to include so many outside stakeholders, individuals, and institutions. Most folks don't know just how much the military does to protect the biodiversity heritage of the United States. The military's full support behind DoD PARC to benefit both the amphibians and reptiles on military bases, while also protecting its mission, is a tremendous success story. I am grateful to be in this position and thank all the hardworking military employees and their contractors who help in the same fashion each day.

Finally, I would like to acknowledge the dedicated men and women who have in the past or now serve in the US military. Many of our peers and colleagues in herpetology have done so, some of whom were inspired to become herpetologists

The author and a Coachwhip (slithering at attention in proper military fashion) at Fort Carson in May 2022. They are both shown in situ, taking part in a herp inventory, directly implementing inventory and monitoring efforts as called out in the Strategic Plan for Amphibian and Reptile Conservation and Management on Department of Defense Lands. The Coachwhip pictured was a solid data point for the installation as part of Fort Carson's natural resource and herpetofauna management efforts. Photo courtesy of Richard R. Riddle.

while serving abroad. Their stories and sacrifices would take many more pages to share, and their legacy lives large.

About the Author

After graduating from Loma Linda University with his PhD in 2009, **Robert Lovich** became a senior natural resource specialist for the US Navy in San Diego, California. His primary duties include managing herpetological and natural resource projects on military lands throughout the United States. Robert has been studying amphibians and reptiles professionally for 25 years. He is the national technical director for the Department of Defense Partners in Amphibian and Reptile Conservation (DoD PARC) since its formation in 2009. Robert has authored many herpetological peer-reviewed publications, popular interest articles, book chapters, and books. When not surfing, golfing, or working on his 1960s muscle cars, his work and interests include natural history and evolution of herpetofauna; application of endangered species conservation; and integrating science into natural resource management.

Part IV

MISHAPS AND MISADVENTURES

28

CLOSE ENCOUNTERS OF THE GATOR KIND

Whit Gibbons

The photographer said, "You might want to wipe the blood off its teeth."

Apparently, my less than collegial relationship with alligators had not ended. The reptile in hand was for a color photo closeup for a magazine article about my putative expertise with reptiles. I had been looking at the reporter as I was rattling on about how experienced I was in conducting research on alligators. As I reached into the bucket, I misjudged how far a three footer could jump up and bite me on the hand. Baby alligators are protected by their mothers, but this one was old enough to take care of itself.

I had the wherewithal to hold the gator with my uninjured hand for the photo but was not aware of my bright red blood dripping from the pointed white teeth of my attacker as it smiled for the camera. But then again, I was well accustomed to being on the receiving end of alligator "back-atcha" behavior, so I was no less dismayed than I had been in previous ignoble encounters with the country's largest freshwater reptile.

My somewhat tumultuous experiences with alligators had started years earlier with the first one I ever caught. I wanted to take the captured gator to the University of Georgia's Savannah River Ecology Laboratory (SREL) to measure it, excited about my first alligator data point. Transporting a 6-ft alligator by tethering it in a pickup truck bed with a rope around its neck had seemed like a reasonable plan. A graduate student and I had passed the rope through the rolled-down windows of the cab, making it impossible to open the doors. Unfortunately, as the gator shifted in the back of the truck, it moved slowly but inexorably toward the driver's side window. To my further dismay, I realized that the rope was moving,

A mother American Alligator defends her nest and eggs with open mouth and loud hissing. Photo by Thomas Rainwater, Randeep Singh, Clarissa Tuten.

and the monster began crawling, mouth open, through the window. I reached our lab's parking lot, slammed on the brakes, and frantically followed the student out the passenger side window. I had some explaining to do to my colleagues who watched, also open mouthed.

As more alligators entered my life, ecological and behavioral research findings drifted away when scientific projects based on robust, quantitative data sets became overshadowed by a single data point—an encounter with America's endemic crocodilian. These single observations continued, until I was eventually able to make a categorical statement: a determined alligator is a worthy opponent that can humble any herpetologist. Most of my research was on amphibians and reptiles other than alligators. But the latter holds the franchise in many southeastern wetlands I have had occasion to enter, leading to too many less-than-ideal encounters of the gator kind.

During one wildlife survey for a resort facility on a coastal island, our lodging was alongside a freshwater lake where a mother alligator lived with her babies. We saw her every day as she patrolled the shoreline. She saw us as we stood on our deck. A cordial relationship was brewing, or so it seemed. As added entertainment we occasionally saw the maternal instinct of the large reptile on display, leading to appreciation and understanding of the close evolutionary relationship between birds and crocodilians. Our house overlooked a golf course, and people routinely hit their dimpled white balls close to the lake that the mother and her babies called home. We smiled when we watched her emerge from the lake to chase a golfer away. I once saw her eat a golf ball. The golfer also observed her choice of snack when he looked back over his shoulder during his retreat. Alligators ingest rocks, pinecones, and apparently golf balls. They retain these in the gizzard, the way a bird does, for grinding their food, which is swallowed whole.

Our relationship with the gator family turned sour when we were asked to catch and remove the mother. Someone considered our reptilian neighbor a pest and had reported her to the resort's management. Rather than viewing the request as an assault against the natural order of the world, I looked on it as an opportunity to assist with environmental preservation. The resort had contacted the state department of natural resources, which had given permission for the gator to be killed as a nuisance animal. However, they also approved the option that would allow us to remove her. We would transport the mother to SREL for our environmental education program. The babies were old enough to survive on their own as orphans.

American Alligators (*Alligator mississippiensis*) build aboveground nests, mounds of mud and vegetation, where they lay 30 or more large white eggs. The decomposing nesting material generates heat as the eggs incubate. The mother typically remains in the vicinity of the nest until late summer when the young hatch. Before then, any trespasser might suddenly find an angry, hissing reptile charging overland. Supposedly, if one backs away and does not molest the nest or dig up an egg, the mother will retreat. Supposedly. "Fool's errand" describes that field experiment. The mother's protective behavior continues after the babies hatch and enter the water.

American Alligators have some notorious people-eating kin in Africa and Australia and have never been completely above suspicion themselves. Thus, they are viewed warily by many visitors to southern swamps and lake margins. A 7-ft behemoth erupting from the water, toothy mouth agape, is a surefire attention getter for golfers, joggers, or investigating herpetologists. Any admiration people might have for a gator's protective maternal behavior is tempered when she defends her offspring from a human. American Alligators, which are shy and peaceful if unmolested, thus become victims of public misunderstanding and undeserved mistrust.

Arrangements were made to capture the alligator we had been observing in the lake. I took three people with me—two of my SREL colleagues, Jeff Lovich and Tony Mills, and my 12-year-old son, Michael. A sure time to locate alligators is after dark. Their reflective red, orange, or yellow eyeshine is unmistakable in the beam of a flashlight.

Scattered ground fog and a slowly rising mist turned the early autumn evening into the setting for a gothic novel. Pine trees splintered the silver light of a full moon, washing the palmetto-lined fairways with shifting slivers of black, white, and gray. As we moved quietly through the gloom toward our quarry, we were serenaded by a chorus of Green Treefrogs, punctuated by the territorial declaration of a Barred Owl.

As we drew closer to the lake, my confidence morphed into smugness knowing Tony and Jeff were with me. Both were large men with experience handling bull gators up to 12 ft long. I anticipated no problem with a 7-ft female. I had brought Michael along to show him how professional alligator catchers such as ourselves accomplish the job.

When we got to the edge of the water, our flashlights picked up eyeshine. Circling the large, ruby red eyes were what appeared to be a swarm of lightning bugs rippling the water's surface. The source was at least 30 pairs of little yellow eyes, babies staying close to their mother.

Our plan was a simple one, a yawner for experienced alligator catchers: we had a plastic-encased cable with a noose, the free end being attached to an 8-ft bamboo pole. When the mother approached shore, one of us would lunge forward and slip the noose over her head. Once the cable was pulled tight around her neck, the three of us would pull her from the water and subdue her enough to put big rubber bands around her snout. With her mouth securely closed, we would be able to carry her to our pickup truck.

Her plan was also a simple one: sit in the middle of the lake surrounded by her babies and pay no attention to the primates on shore.

Our revised plan: see if one of the babies was close enough to shore to catch. It would give a distress call to alert its mother. Upon hearing the distinctive grunting sound, the mother would come to the edge of the lake to investigate. We could then carry out our original plan.

We caught an adventurous hatchling in vegetation along the shore. All of us—four humans and one mother alligator—heard the yelping sound of a frightened baby. To my satisfaction, here came the mother. Two crimson eyes rapidly closed the distance between midlake and shore. I handed the baby to Michael, and the rest of us jumped behind two big pine trees. We were out of sight of the approaching mother. I told Michael to stand in plain view about 20 ft away from us.

I knew the plan was working when the mother was only a few feet from the shoreline, and Tony prepared to jump from behind the tree to noose her. I knew the plan was not working when she did not even slow down at the water's edge. The snarling reptile emerged from the bottom sediments, scrambled onto shore, and headed straight toward Michael. The alligator's tail thrashed sweet myrtle bushes along her path and the incongruously fragrant perfume mingled in the air with the musky smell of pond water. She passed between the pine trees as we stood immobile, having been caught completely off guard.

Michael also stood still, holding the baby, and said, "Dad, Dad, what do I do now?"

Coming out of our collective trance, the three of us offered professional guidance.

"Climb a tree!" yelled Jeff.

"Throw the baby in the lake!" Tony shouted.

"Run!" I roared.

Dutifully obeying his father, Michael ran. The last thing I saw as he disappeared into the darkness was the squeaky toy of an alligator he held above his head. An angry alligator is fast for a short distance, but fortunately a scared 12-year-old is a lot faster in a longer sprint.

As thought and mobility returned to the rest of us, Tony caught up with the mother and slipped the noose over her head. Jeff and I grabbed the bamboo pole, bracing ourselves as the cable tightened. Finally, the plan was working. Or was it?

With surprising agility, the alligator reversed direction, ran between Jeff and me, and headed back to the lake. By that time, all three of us had a good grip on the bamboo pole. With our backs to the lake and the pole parallel to the shoreline, we were yanked to the ground, dragged for 10 ft, and then pulled down the slippery slope into the lake.

Fortunately, the noose slipped off before the mother disappeared beneath the surface. Unfortunately, this meant an unrestrained alligator lurked unseen in the black lagoon. We scrambled out faster than we had gone in. If in my haste to reach shore I inadvertently stepped on Jeff and Tony's back and shoulders, I feel certain I apologized to them when we reached dry land.

We put some distance between us and the lake, then stood dripping in silence until Michael emerged from the woods. We returned the baby to the water. If there had been a scoreboard, it would have read "Mother Gator 3, Herpetologists 0."

On our way back to where we were staying, I suggested to Michael that the night's events would be of little interest to his mother, so he need not mention them. After all, as that night illustrated, some mothers can be very protective of their young.

About the Author

Whit Gibbons grew up in Alabama and Louisiana, spending as much time as possible in woods, swamps, and fields looking for reptiles and amphibians. A master's degree in biology (University of Alabama) was a natural outgrowth of those childhood pursuits. Getting a PhD in zoology (Michigan State

University) set him on a career he pursued for 50-plus years: research ecologist with a particular fondness for herpetofauna. Most of his time was at the University of Georgia's Savannah River Ecology Laboratory, with sabbaticals at the University of Michigan and the Smithsonian Institution's National Museum of Natural History.

29

DON'T TREAD ON HER

Kelly Zamudio

I felt a squishiness under my hiking boot, almost like I had stepped on a fresh cow patty, and then immediately heard the explosive rattling of an angry Black-Tailed Rattlesnake under my feet. I looked down at my left shin and could clearly see two puncture marks, little over an inch apart. As the annoyed large female *Crotalus molossus* back-circled twice and then crawled off rattling my first thought was "Well, I've always wondered what this would feel like."

August 8, 2002 was the day my life and the life of that rattlesnake collided ever so briefly. I was a new assistant professor at Cornell University, and I was collecting in Skeleton Canyon, in the Peloncillo Mountains of southwestern New Mexico. My goal that day was to collect two species of lizards that were part of a study on genetic diversity across the southwestern sky islands, a series of beautiful mountain ranges that rise above the desert in that part of the world. I had worked that canyon a few times before, sometimes even hiking alone, carrying my lizard lasso, scanning the rock walls for lizards. That particular morning, fortunately, I was accompanied by my first doctoral student, Lauren Chan, and we were just hiking into the canyon when the snakebite happened.

I had been working in the sky islands of southern Arizona and New Mexico for a few years, and my husband Harry Greene and our colleague David Hardy, a Tucson anesthesiologist, had a long-term study on Black-Tailed Rattlesnakes in the neighboring Chiricahua Mountains, so I was familiar with these docile serpents. This species is so calm it is rarely involved in snakebite incidents. To be bitten by one you really have to be unlucky enough to step on it, and that was just not my lucky day. In retrospect I see how this happened. It was morning, and the sun had just crested over the canyon walls, creating dappled shadows on the yellow and brown oak leaf litter that lined the ground—a perfect match to the yellow and brown diamonds of blacktails. Perhaps she had been coiled there all night and was just warming up in the morning sun to start her day.

What happened next was a chaotic, yet coordinated, blur. Within seconds I felt an odd sensation in my joints and a distinct burning sensation at the site of the punctures, an indication that this had not been a dry bite. David Hardy, who was interested in rattlesnakes from both natural history and public health aspects, always emphasized the need for field biologists to have a snake bite protocol in place and, especially, to know the location of the nearest hospital that stocked antivenom. So, according to our plan, Lauren and I headed to the clinic in Douglas, Arizona, about an hour away. The pain in my leg was getting steadily worse, and it took all my mental energy to keep it at bay. The sensation was of burning, like I was being branded, but the iron tool never left my skin. I could feel the front of tissue-destructive venom creeping up my leg, and I imagined a band around my knee and willed the pain to stay below it. Of course, no willing in the world keeps venom from doing exactly what it evolved to do, which is to break down cells and spread through tissues. But my mental exercise helped me manage the pain and keep us as calm as possible, despite that adrenaline was rushing through our bodies. Lauren was flying down the highway to Douglas, driving over 90 miles an hour, and I remember looking over at her and seeing a bead of perspiration running down her forehead.

At the Douglas clinic I allowed myself to feel relieved, maybe everything was going to be okay. They immediately put me on morphine to manage the pain, and after convincing the doctor that I was sure I was bit by a rattlesnake (*and yes, I am SURE it was not a coral snake*), he started my first vials of antivenom, delivered intravenously. I was airlifted to a Tucson hospital in a helicopter that was mostly windows, with a nurse by my side. That flight over the desert flats, with the sky islands jutting up below me, while on morphine, was one of the most spectacular experiences I have ever had. The clinic had contacted Harry, and he and David Hardy would meet me at the hospital in Tucson. I was safe now.

Then began the roller coaster. Mine was not a simple envenomation, and several factors contributed to a serious condition that eventually required surgical intervention (Hardy and Zamudio 2006). Because this was a defensive bite, the fangs penetrated well into the muscle bundle on my shin and delivered venom deep within. I received antivenom treatment, which slowed tissue destruction, but the venom of blacktails is not included in the cocktail of venoms used to produce a polyvalent antivenom. In theory, this antivenom is effective for rattlesnake bites by all species but will differ in efficiency across species, especially if some have unique components to their venom. Another telltale sign that my envenomation was unusually deep is that I never developed the skin blebs and blisters that are typical of rattlesnake bites. By the time my antivenom course was completed, it was clear that the amount of tissue damage in my shin was extensive and, without surgical intervention, it would compromise the entire muscle bundle and possibly my lower leg. Three days after my bite I underwent an emergency fasciotomy to relieve the pressure in the anterior compartment of my lower leg caused by the large amount of necrotic tissue and inflammation.

Over the next two years I worked hard to regain my mobility. The surgeon who did the fasciotomy told me that I should expect to limp for the rest of my life. I thought to myself, "You are just an MD, you might know about means, but you don't really understand variance." My husband, who knows me best, has said that was the best prescription the doctor could have given me. Nothing like some spite for motivation! A month after my accident, back in Ithaca, I started the long process of building up the damaged shin muscle, the one that lifts your foot up toward your shin (the technical name of this muscle is *tibialis anterior*). Excessive damage to this muscle causes "foot drop," such that your toes do not clear the ground and requires compensation by elevating the knee. I knew that to have a normal gait I needed to reach that magical neutral point, and making that mark became my obsession. I did everything I could to build up that muscle that had been so damaged by a big load of venom. Approximately a year into my recovery my progress began to slow because of the formation of scar tissue that was binding the tendons and muscles in my shin. One possible treatment is scar excision, a surgical procedure that removes the scar tissue that hinders movement. This procedure can potentially have diminishing returns because, of course, any surgical intervention also creates scar tissue. This time, with no emergency decisions being made, I did my homework, read about the procedure and its likelihood of success in my case, and interviewed multiple surgeons to gauge their experience before selecting a young doctor in New York City. That surgery increased the mobility in my toes and got me a few more degrees of upward foot movement, past the critical neutral point.

I improved enough to be able to walk with a normal gait and return to hiking, running, and other activities. I made it a goal to run a recovery marathon (because why not?), and with the help of many running friends I trained on the running trails in the state forests around Ithaca. I ran my first marathon in Mexico City with my good friend Gabriela Parra, just a few years after my accident. We then did some repeats the following years, in Chicago and Washington, DC. It took many years and uncountable hours of work, and each time I crossed a finish line I imagined that little muscle getting stronger and stronger. Now the 11-in fasciotomy scar on my shin has faded, and it is just a part of me. Most of the time I even forget it is there.

Fieldwork will always involve danger. Especially when working in remote places, any number of accidents can happen and turn a fun and productive scientific endeavor into a life-threatening situation. Venomous snakebite accidents, although very rare, are among those. I have always had a healthy respect for venomous snakes, but I never feared them. I have always told my students that if we take the right precautions we can reduce the probability of animal-induced accidents, and the largest threat to us in the field is likely to be interactions with other humans. To this day I believe that is true. I have been the most scared in the field when we happened upon humans, sometimes even armed, and did not yet know their intent. Being isolated, with no immediate escape, and with no one to turn to for help is especially difficult for women in the field. That feeling of vulnerability is no fun at all. In contrast, most of my encounters with venomous snakes in the field have included only the excitement of spotting them, admiring their beautiful patterns and scales, and marveling at how they make a living.

After my accident, I wondered whether my relationship with venomous snakes would change. Would I fear another encounter so much that I would not enjoy fieldwork anymore? Would I blame rattlesnakes for what I went through? Some

A Black-Tailed Rattlesnake (*Crotalus molossus*) in a resting coil. Huachuca Mountains, Arizona, August 2020. Photo courtesy of Jeff Martineau.

things did change. The first couple years after the bite I was hyperaware of movement on the ground, even if just a hopping insect or blades of grass blowing in a breeze. My brain was hypervigilant. If twigs or stems brushed up against my lower legs I would start and immediately look down. It took a few years for these reactions to pass. I still dislike walking through high vegetation where I cannot see my feet, even when wearing snake chaps. But I adapted and have since had great field experiences in snaky places and many safe encounters with rattlesnakes, jararacas, and other pit vipers—back to our relationship of admiration and healthy respect.

I have thought often about that female blacktail in the last 20 years. Rattlesnakes are long-lived, and it is conceivable that she still lives in that same canyon, still wakes up every morning to the warmth of the early morning sun rays, just as she did that day. I wonder if she had litters every couple of years and how many times she unobtrusively watched hikers go by, undetected. I hope she did not get killed with a shovel or machete by someone after a rattle trophy. I hold no grudges and certainly do not blame her for protecting herself. We share something in common. If I were tread on that way, you can be sure I would have bitten too!

Reference

Hardy, D.L., and K.R. Zamudio. 2006. Compartment syndrome, fasciotomy, and neuropathy after a rattlesnake envenomation: Aspects of monitoring and diagnosis. *Wilderness and Environmental Medicine* 17:36–40.

About the Author

Kelly Zamudio was born in California and early on her family immigrated to Brazil, where she was raised. Kelly received her BA from University of California (UC)–Berkeley (1991) and her PhD from University of Washington (1996); she was a postdoctoral fellow at the Museum of Vertebrate Zoology at UC Berkeley (1997–1999). Kelly was a professor at Cornell University from 1999–2021 before moving to her current position as a professor and Endowed Fellow of the Doherty Regents Chair in Integrative Biology at the University of Texas at Austin. She is an elected member of the American Academy of Arts and Sciences. Her research focuses on the processes leading to the origin and maintenance of diversification in reptiles and amphibians. Kelly and her husband Harry Greene share a ranch in the Texas Hill Country with some pet longhorns and numerous Western Diamondbacks.

30

A SNAKE TO DIE FOR

Robert W. Hansen

Most Americans had no idea what was going on in Sinaloa, Mexico, in the 1970s. The fertile soils in the western foothills of the Sierra Madre Occidental provided an ideal setting for growing tomatoes destined for American salads, but also opium poppies and marijuana, and these operations were run by Mexican drug lords. The "narco violence" we now know about all too well thanks to Netflix was there but on a much quieter scale—and the murders on the streets of Culiacán (Sinaloa's capital and the epicenter of the West Coast drug trade) rarely made headlines in Los Angeles. Back then we had no idea about any of these hazards— we just wanted to find snakes. Lots of snakes. Snakes we couldn't see in California or Arizona or Texas.

Ron Tremper, my closest herping buddy back then, and I made our first pilgrimage to northwestern Mexico in the summer of 1975. I was a college undergraduate and Ron had just completed his bachelor's degree in zoology, both of us at California State University–Fresno. Ron was the first "serious" herp person I'd met. He'd already done a Peace Corps stint in Cameroon and worked as a summer intern at the Arizona–Sonora Desert Museum in Tucson. As a teenager, he'd corresponded with the famous rattlesnake expert Laurence Klauber, just missing a personal meeting with him as Klauber fell ill and died soon after.

Ron and I first met in September 1974. We lived in an area that seemed devoid of other aspiring herpetologists, unlike the major metro areas of California—San Francisco, Los Angeles, and San Diego. A regional herp group did not exist. There was at least one other local herp guy we knew of, but we suspected he was a poacher and smuggler, so he was off limits. I think both of us were excited to have met a kindred spirit, somebody who was happy to "talk herps" for hours on end. We each had jobs at a local mall and coordinated our work breaks. It was during one of these breaks that Ron hatched the idea of going to Mexico. Mexico! The thought of going to Mexico to hunt snakes was pretty wild (maybe even dangerous?), but Ron was determined that we should go. And so the planning began.

Ron took the lead in obtaining a scientific collecting permit from the Mexican government, a months-long process but ultimately successful. He had a new job at the Fresno Zoo and would later become its first curator of reptiles. In the meantime, a state-of-the-art reptile house was to be constructed, and we hoped to obtain some of the first display animals that would eventually live there.

I spent months in the university library researching species distributions and activity periods. There were no field guides for Mexico available back then. Color photos for many species had never been published. I filled three large binders with photocopies of scientific papers concerning the herpetofauna of northwestern Mexico and made regional checklists for the places we planned to hunt.

I owned a near-new VW Beetle, and we decided that it would be a reliable and economical vehicle for our trip. Ron had expertise in fabricating stuff, so he built a large metal storage locker that would fit on the car's roof rack. After reading one-too-many travel warnings about what to avoid eating and drinking while in Mexico, we decided we should take our own food and would prepare our own meals. Our menu was not too exciting—Spam, canned corn—there must have been other stuff, but that's all that I can remember decades later!

July 13. Finally, the day of departure arrived. The first leg of our epic adventure was the long drive from Fresno, California, to Tucson, Arizona. We made it in time for a bit of road cruising in the Gates Pass area on the outskirts of Tucson, finding a California Kingsnake and a Black-Tailed Rattlesnake. The next day we crossed the US–Mexico border at Nogales and continued south to Hermosillo, Sonora. After securing a motel room and enjoying a meal of Spam and corn (!), we headed out for a first night of hunting in Mexico. We were not disappointed. From 8:40 p.m. to 1:55 a.m., we found 16 snakes. This was Upland Sonoran Desert, thus many of the species were familiar to us because they ranged into southern Arizona. Our highlights were a beautiful Sonoran Coralsnake and a Tiger Rattlesnake.

We spent a second night in the Hermosillo area, but by then the subtropics were practically crying out to us. On the next morning we drove south on Highway 15, headed for the small town of Alamos, tucked into the foothills of the Sierra Madre in southern Sonora. Alamos had long been a popular destination for herpetologists, bird watchers, and naturalists of all stripes. This was near the northern limits of the subtropical thornscrub forest that cloaked the Pacific slopes of the northern Sierra Madre. The regional checklist I'd assembled suggested it was possible to find such novelties as Mexican West Coast Rattle-snakes, Spotted Cat-Eyed Snakes, Cantils, and Beaded Lizards, among others. Species diversity here was notably greater given that this was the meeting place of two major biotic communities—Sonoran Desert Scrub in the lowlands and Sinaloan thornscrub in the foothills—producing a curious mix of familiar desert regulars like Sonoran Gophersnakes alongside Boa Constrictors!

Alamos met and exceeded our expectations. The allure of road hunting for snakes at night is that you can cover a lot of territory in a short time. And, you don't know what you might find next, maybe even a species you've never seen before. Although we were able to identify nearly everything we encountered, one Alamos snake was a puzzle. It was a dark snake with cream-colored bands, smooth and shiny body scales, a blunt tail and measured about 450 mm in total length. We had no idea! (Upon our return to the United States at trip's end, we stopped in at the Arizona–Sonora Desert Museum, where resident herpetologists Merritt Keasey and Steve Prchal readily identified our mystery snake as a *Sympholis lippiens*, sometimes called the Mexican Short-Tailed Snake).

It was time to venture farther south, into Sinaloa, where new landscapes and new species beckoned. We checked into the Motel York in the small town of Guamúchil, on the coastal plain about 100 km north of Culiacán, Sinaloa's capital. Our trip fell during the summer monsoons—a hot and humid period punctuated by near-daily afternoon rain showers. Night driving followed, usually revealing uncounted frogs and toads, along with a nice diversity of snakes. Hunt late into the night, sleep in the next day. That was our routine. On this particular day (24 July) we lounged around the motel pool under threatening skies and hot, humid conditions. The characteristic chatter of the ever-present Boat-Tailed Grackles filled the air. Soon enough, the skies opened, and we headed for cover. However, the storm was short-lived, and the sun made an appearance by late afternoon. This cycle—rain showers followed by sunny skies—usually meant that snakes would be on the move that night.

We studied our Sinaloa map, hoping to find roads beyond the ever-expanding maw of industrial agriculture, where native habitats were likely to produce snakes. On the map we noticed a paved road running from the Pacific lowlands winding into the foothills of the Sierra Madre. The "Badiraguato Road" looked promising and would be one of our targets that night.

We both had species we hoped to find on this trip. I was particularly interested in *Loxocemus bicolor*, sometimes referred to as the Mexican Burrowing Python, at the time considered the only New World member of the Pythonidae (pythons) but really more of an evolutionary mystery. Over the years it has been kicked around like a phylogenetic football, variously grouped with pythons or boas, eventually being placed in its own family, Loxocemidae. I badly wanted to find this enigmatic snake! It does not range as far north as Sinaloa but would be sought later in our trip. Ron had his sights set on finding a tropical milksnake (aka Sinaloan Milksnake, then known scientifically as *Lampropeltis triangulum nelsoni*). This beautiful red, white, and black banded snake was one of many subspecies then placed within *L. triangulum*, a species with a range that extended from Canada to Ecuador. Ron had long been interested in kingsnakes (*Lampropeltis*), and this would be a chance to add a spectacular new species to the list of those he'd seen in the wild.

Before leaving California, we'd been warned not to drink the water, to avoid unwashed fruits and vegetables, not to keep all our cash in one place, and not to be surprised by the widespread practice of *mordida* (small bribes solicited by local police or other officials). But nobody told us about the Narcos.

Once darkness arrived, we headed out from the hotel parking lot, turning south onto the two-lane Highway 15—a crowded mix of buses, trucks, and rain-filled potholes whose depths could not be gauged. After hunting some side roads, we eventually reached our turnoff to Badiraguato. I was piloting our trusty VW that so far had held up well on Mexico's roads. According to my field journal, it was 10:20 p.m. and the air temperature was 80 °F under overcast skies. In the distance we saw some faint lights on the other side of a one-lane wooden plank bridge over a small stream. Onto the bridge we drove—it was quite narrow so I was creeping along.

As we progressed, we noticed a group of men who appeared to have automatic weapons and pistols—all pointed at us. We did not have the option of turning around so we moved forward at about 10 mph. As the distance between our car and the men closed, the brilliant red bands of our target snake materialized in front of us . . . right on the bridge! Ron screamed "*nelsoni!*" and without concern for the consequences, flung open the passenger door and rushed forward to capture the snake—and in doing so running right at the gunmen, who were only a few meters behind the snake.

I was already mentally processing how I was going to tell Ron's wife that he'd gone down in a hail of gunfire . . . for a snake. Meanwhile, in an instant Ron had captured this brilliant meter-long tropical milksnake—a dead ringer for the local and highly venomous Mexican West Coast Coralsnake. In typical milksnake fashion, it sank its teeth into Ron's hand and started chewing, quickly drawing blood. Still no gunfire. Ron held his prize aloft, unconcerned about the bites, blood streaming down his arm, all brightly lit by the headlights of our car. He is overjoyed, I am incredulous. Still no gunfire. Meanwhile, the Mexicans are now very agitated, yelling at Ron, "*coralillo!*" (Spanish for coral snake). We make it across the bridge and are pulled from the vehicle at gunpoint. The Mexicans can't believe Ron isn't affected by the bite of the snake. Using my limited Spanish, I tell them that we are American students looking for snakes. They think we are crazy. *Gringos locos* they called us. After checking our car and finding only more snakes (in bags), they actually let us continue hunting on their road.

Years later, I learned that the Badiraguato region was not just another random drug-growing area in the foothills of the Sierra Madre. It holds special status in Sinaloan drug lore as the birthplace of Joaquín "El Chapo" Guzmán, the notorious Mexican drug lord. It was here that his father, a *gomero* (opium poppy farmer), taught his son how to grow, harvest, and transport their illicit crops.

A snake to die for—almost.

About the Author

Robert W. Hansen has a long-standing interest in the amphibians and reptiles of the American Southwest and Mexico. He served as editor of *Herpetological Review*, a quarterly journal published by the Society for the Study of Amphibians and Reptiles (SSAR), from 1991 to 2021. In 2015, he received the SSAR Presidential Award for Lifetime Achievement in Herpetology. An accomplished photographer, his photos of herpetological subjects have appeared in numerous books and journal articles. Currently, he is an affiliated researcher at the Museum of Vertebrate Zoology, University of California–Berkeley.

31

GOOSE ON THE ROAD

William W. Lamar

Mexico is Mecca for biologists. Blessed with multiple seacoasts, jagged mountains, sweeping deserts, roiling rivers, and stunning forests, the country boasts an unparalleled richness of plants and animals. That extends to reptiles and amphibians, which led, some 40 years ago, to the sawing and hammering in Jonathan Campbell's Arlington, Texas, driveway. We were sprucing up his redoubtable Ford F-150 pickup, known affectionately as The Blue Goose, by retrofitting the truck bed and camper shell with the finest cheap plywood available. Things had to be arranged so that three of us could sleep in the tiny space yet carry collecting equipment, canned goods, and a bulky cryogenic storage dewar needed for hauling liquid nitrogen to preserve tissues for biochemical studies. Crowded but functional, the camper endured us for over 20,000 mi, crisscrossing Mexico and spawning a book's worth of gritty stories best suited for campfire airing. The Blue Goose was stalwart, as evidenced by the never-ending quantity of fur distributed throughout the motor and dating from a violently unfortunate kerfuffle with a black bull that galloped, during dark of the moon, onto an unlit Mexican road. The bull lost.

After jumping through interminable paperwork hoops to obtain the University of Texas at Arlington museum's collecting permits, we hauled our ambitious herpetological Easter egg hunt south for the summer. Our interests embraced all things Mexican with some priorities. Dave Hillis sought specimens and tissue samples of anurans traditionally known as Leopard Frogs. His dissertation revealed that what was thought to be *Rana pipiens* was actually a diverse complex of species. This threw the status of innumerable lab-based studies on physiology into turmoil because no one knew what kind of frog had been used. Arguments about the proper relationships among these amphibians, perhaps in the genus *Lithobates*, continue to this day, but Dave's research opened the door. Jon, an unparalleled authority on Mexican and Guatemalan herpetofauna, likely already had an eye on producing a scholarly treatise covering the country; it's a staggering task with which he is currently engaged. Having recently entered grad school after delightful years living in Colombia, I was interested in everything and grateful to be getting back into the tropics. Time in the field teaches valuable lessons, but I hoped it would also ease the surprisingly intense culture shock that wracked me upon my return to the United States.

Happily, our agenda was loose so we could indulge a collective passion: getting into Mexico's inaccessible areas, mostly upland forests—like the Sierras Mixe, Juárez, and Zongolica—and plumbing their depths for zoological novelties. We barreled across the country, breezing through dusty hamlets, dodging insane truck drivers while snatching rattlesnakes like noodles off the roads. Revving and braking The Blue Goose—at times with considerable artistry as it lacked four-wheel drive—we got ourselves into places any popular book would call "forgotten Mexico." I'm one of the few Acapulco visitors who have never seen the beaches; we simply dipped down into the city's outskirts, gorged on street food, and sought other roads leading back into the highlands. The Blue Goose did not lack idiosyncrasies. On Sundays in rural Mexico, any car mechanic worth his salt was inebriated; enlisting their help required participation. More than once, after obligatory bouts with pulque, mescal, or tequila, the truck was restored to working order but our hangovers were horrifying. Other circumstances called for celebration. After amassing a collection, we conducted a marathon session of specimen photography, tissue extraction, tagging, cataloging, and preservation before continuing to our next destination. Benefitting from the devalued Mexican peso, we rewarded ourselves with excellent local food and libation, at times boisterously consuming so much of both that the antagonized help must have longed to serve us using sticks. Life was good.

Passing buildings adorned with exquisite Talavera ceramic tiles in venerable Puebla, we made our way south toward Oaxaca, steering through colorfully named hamlets like San Salvador Huixcolotla, Tecamachalco, and Tlacotepec de Benito Juárez. In the environs surrounding San Andrés Cacaloapan, we sought *Mixcoatlus melanurus*, the Black-Tailed Horned Pitviper. Our efforts turned up only Mexican Pygmy Rattlesnakes (*Crotalus ravus*) and alligator lizards (*Gerrhonotus* sp.). Departing the next morning we rolled into Tehuacán, where we had a choice. We could angle southwestward toward Santiago Chazumba and a town with the commanding name of Heroica Ciudad de Huajuapan de León, one of countless places in Mexico where fabulous dining awaited. That might include tacos made from *nopal* (cactus pad) tortillas laden with refried beans and *chapulines* (toasted grasshoppers) ... delicious! We'd pass through the Zapotitlán Salinas basin, an otherworldly, cactus-studded, rain shadow rife with endemic reptiles including Pueblan Milksnakes (*Lampropeltis polyzona*), Mexican Lyresnakes (*Trimorphodon tau*), and Giant Blindsnakes (*Rena maxima*). Plus, we could hit the adjacent Sierra Acatepec for those Black-Tailed Horned Pitvipers.

Off to our left, in northeastern Oaxaca, loomed the imposing Sierra Mazateca, studded with karst peaks and bedecked with pristine cloud, temperate, and tropical rain forests. We opted for that route, turning east at Teotitlán de Flores Magón onto a surprisingly (even for Mexico) inhospitable road that wound insanely up to Huautla de Jiménez. The indigenous Mazatecan culture there embraces spiritual use of entheogens derived from hallucinogenic mushrooms (*teonanácatl*) and *Salvia* (*xka pastora*). After seminal anthropological and ethnobotanical investigations in the 1930s and 1940s, things exploded in 1957 when *Life* magazine published Gordon Wassen's self-aggrandizing account "Seeking the Magic Mushroom." Owing to intense interest in the psilocybin-containing fungi, the area has, for decades, attracted a global amalgam of tourists, scholars, and famous dilettantes, including Bob Dylan and John Lennon, all undoubtedly dizzy after negotiating that road.

We were passingly acquainted with the region's psychedelic history, so when a glassy-eyed, mumbling, traditionally clad Mazatec Indian carrying a shoulder bag crammed full of mushrooms appeared out of nowhere to stare at us while waving a shotgun skyward we found it only moderately disconcerting. He seemed to be under the influence of something but let us pass without incident. A cursory glance at a local topographic map revealed extremely compressed contours, especially east of Huautla, where things got interesting. The squiggly red line indicating a road on our map abruptly ceased, and a blank spot intervened before things picked back up at the town of San Felipe Jalapa de Díaz, 40 mi to the east. There was a gap in the map.

Normal people avoid such spots, or at least have a noble cause that merits heading into one, but we were still young enough to possess more brawn than brains, and we were not normal. Not all road trips require roads, right? That made perfect sense to us, so in a display of hubris best described as flamboyant, we blithely piloted The Blue Goose into the void; it's technically called the Huautla Fault. Mexican geologist Manuel Villada wrote of the Sierra Mazateca in 1906: "As one advances, an interminable series of peaks appear, of different altitudes and shapes, simulating the waves of an agitated ocean." When the good lord made Mexico, I'm guessing the acreage was apportioned up the Pacific Coast, across the Baja, Sonoran, and Chihuahuan Deserts, and thence southward along the Gulf shoreline. Excess land accrued as things approached the narrow Isthmus of Tehuantepec, so at the confluence of Veracruz, Puebla, and Oaxaca the creator simply wadded it up and planted it where the Trans-Mexican Volcanic Belt crashes into the southern Sierra Madre Oriental Mountains. Iron the place flat and you'd have another country. The descriptor "craggy" is inadequate.

On one particularly steep incline we blew a tire and had a grand time fixing it, which meant we now lacked a spare. Later The Blue Goose rolled through Nubiñuutun, a grassy, incredibly isolated Mazatec village. We narrowly missed flattening a crying little girl's pet chicken and bewildered the scrambling locals, shocked to see a vehicle and seedy-looking foreigners. More than anything we encountered, that cloistered place drove home what was meant by a gap in the map. We drove into nowhere and also, it seemed, back in time. As we braked into the bowels of the gorge, the oatmeal grays and tans of the highlands yielded to shades of green. A tentative route materialized, which took us in a near free-fall for 30 mi, zigzagging some 5000 ft in elevation from cold, damp Huautla down to the limpid and beautiful Río Uluápam, which had—astonishingly—a decent bridge and improved driving conditions on the far side. During most of our descent, we saw no one, but that bridge meant something lay ahead. It was dusk, and we could already hear the distinctive "WRAACK" of unknown calling frogs; we stopped to collect.

Some 6,500 ft above us lay Cerro Rabón, a vast Cretaceous limestone plateau punctuated by deep arroyos and valleys. Surface runoff and rainwater percolates through its porous karst, creating caves and waterfalls; indeed, the deepest cavern known in the Western Hemisphere lies in the Sierra Mazateca. Upstream from the bridge, the Uluápam emerges from a mountainside cave in a beautiful cascade. The river was as clear as a swimming pool, meandering among massive boulders and bordered by mesophyll forest. Scattering, we tracked frog calls, finding Schultze's Streamfrog (*Ptychohyla leonhardschultzei*), Small-Eared Treefrogs (*Rheohyla miotympanum*), and ubiquitous Mesoamerican Cane Toads (*Rhinella horribilis*). As darkness settled in, explosive breeding choruses of Mexican Treefrogs (*Smilisca baudinii*) seemed to laugh at us "HA-HA-HA!"

Flush with success, having conquered the worst stretch of the phantom road and discovered a frog-filled oasis, we had the last laugh, loaded our catch into The Blue Goose, and headed east. The narrow path (hardly a road) abutted the Santo Domingo River valley, and forest brushed against the sides of the truck. Entering a curve, The Blue Goose abruptly stopped moving and died. It was as though a giant hand had detained us, replacing our cheery progress with silence. We feared a failed universal joint or worse. Dave, seated by the window, started to step out. Making the understatement of the century, he said, "Uh, we should get out on the other side." He had opened the door to stare down into black emptiness.

Part of the road, over an arroyo concealed by vegetation, had dropped from under us, leaving The Blue Goose beached and delicately balanced, the right rear tire dangling over the precipice. Quickly and carefully, we unloaded everything; the truck could somersault into the gorge at any moment. We sweated in the humid air as clouds of mosquitos descended upon us, overjoyed at the foreign food. And those Mexican Treefrogs erupted raucously "HA-HA-HA!" Our luck, it seemed, had run out.

Flashlights were cumbersome affairs; their incandescent bulbs quickly drained the best batteries. I had invested in a fancy, 6-volt cave light—rechargeable via the truck's cigarette lighter. Since this might last the longest, I went for help while Jon and Dave guarded the truck. Worried about our prospects, I trudged off, leaving my companions to the mosquitos . . . and the triumphant frog laughter. On the plus side, the going was easy. The road continued to drop slightly as I hiked out of the sierra's embrace, passing through groves of sweetgum trees, Mexican sycamore, and *agotope* amidst the riverine profusion of gingers and vines.

I passed rivulets oozing from the higher ground to my left where more Mexican Treefrogs unfailingly burst forth, "HA-HA-HA!" I wondered how long it might take to find help (let alone the cost) at the next town, San Felipe Jalapa de Díaz, 10 mi distant. I fretted more over the likelihood The Blue Goose would take a terminal tumble. An hour later my vaunted cave light dimmed, so I walked hunched-over to see the ground. It went out. Guided by the stars, I staggered on, taking the occasional spill for my troubles. Parched, I felt my way into the foliage and crept toward the sound of a bamboo sluice I knew must be embedded into a trickle; it's the way to make a spigot in the mountains. Feeling for it and hoping my fingers wouldn't encounter a venomous snake, I slaked my thirst.

Periodically I tried my light, gaining a minute of feeble illumination before it died again. Suddenly pitching forward I tumbled broadside over a sleeping cow. There were several, and I crashed squarely into their midst, a sign of humans if I wasn't first trampled to death. I balled up, protecting my head as the bovines staggered to their feet. Miraculously I felt no hooves, and their torrent of excrement missed me. Was my luck changing? I continued, slowly feeling my way along. Astonishingly, three Indians, traditionally clad and swiftly walking abreast in the dark, appeared. I heard strange glottal/nasal intonations that seemed alien. "*Buenas noches,*" I offered in Spanish, and they responded. When I asked for directions they resumed their peculiar conversation and kept walking. They were speaking Jalapan Mazatec.

Dawn's first streak etched the horizon and a light twinkled in the distance; I had reached town. "You need to see don Jorge," a man told me, so we approached a darkened house. Sleep in Latin America isn't sacred, but I was embarrassed rousting the family. Jorge responded loudly and emerged, belly protruding, clad in underpants. My worries at offending him were unfounded, but hopes of economizing faded when he leered predatorily at me and said, "*Buenos días, Meester.*" I talked as he walked down a long green hall. Behind the door at its end sat an elevated, throne-like commode. King Jorge plopped down and, leaving the door open, indulged in his morning evacuation, shamelessly punctuating my testimonial with explosive bursts of flatulence. Feeling like the court jester, I longed for the mosquitos and laughing frogs. Intense bargaining ensued; he had the upper hand, so we settled on an exorbitant sum for his services.

We careened through the empty streets in Jorge's truck as he rounded up assistants, machetes, and car jacks. Looking The Blue Goose over, he barked orders, and a helper gingerly slipped down into the arroyo, actually locating a tiny spot on which to plant a jack with blocks stacked high on top. Gently raising the truck a few inches, we all pulled from the other side, causing it to lurch off the blocks and gain ground. Several more times and we were back on firm terrain. It was a display of Mexican ingenuity at its finest and costliest. I fell into a deep, exhausted sleep as we drove away. My dreams were punctuated by staccato peals of frog laughter "HA-HA-HA!"

Jorge (center) and an assistant work to rescue The Blue Goose, which is teetering on the brink of a precipice; William W. Lamar (left) looks on. Photo courtesy of David M. Hillis.

Epilogue

Two decades later, those stream frogs we collected were determined to be a new species, *Ptychohyla zophodes*; the type locality is the Río Uluápam oasis. Hippies seeking magic mushrooms in Huautla have been replaced by moneyed myco-tourists. The website dangerousroads.org refers to our track, now Mexico's Federal Highway 182, as "a very scenic and dangerous route," that "dizzies into the mountains" and is "not recommended for those who suffer of dizziness or anyone who doesn't do engine brake." All this fuss over a now-paved road? They should've tried it back in the day. HA-HA-HA!

Reference

Villada, M.M. 1906. Breve noticia de un viaje de exploración a la gruta "Nindó-Da-Gé" o cerro del agua crecida. *Anales del Instituto Nacional de Antropología e Historia*, Mexico.

About the Author

William W. Lamar is the retired president of GreenTracks, Inc., an ecotour company specializing in the Amazon basin. A graduate of Rhodes and the University of Texas at Arlington, Bill has lived or worked in 11 Latin American countries; his interests include systematics, natural history, and conservation of Neotropical amphibians and reptiles. He staunchly supports field biology in all its forms. Bill spends as much time as possible in rainforest and still leads tours to the tropics.

32

LOST ON THE PUNA

John E. Simmons

It was dark, near midnight. A thick, swirling fog and steady cold drizzle were soaking through my jacket. I had been looking for frogs long enough that the batteries in my headlamp were dying, so I started back to camp through the wet grass, trying to avoid the mossy puddles. As I trudged along, something didn't seem right. I stopped. I was no longer sure which direction our camp was, and it was too dark to see any landmarks. I was lost.

Getting lost is always unnerving, and in this instance, both an annoyance and an embarrassment as well. The annoyance was knowing that continuing to wander would probably take me further away from camp, so I had to find the driest place I could to spend an uncomfortably cold, wet night. The embarrassment was because I should have been more cautious. I wondered whether my companions, by now asleep, would notice I had not returned and begin to worry. In any case, I knew that they would never let me forget that I got lost.

It was 8 February 1975. I had been traveling through South America with William E. Duellman, Linda Trueb, and their daughter, Dana, for nine months, living in a custom-built, 24-ft camper mounted on a Ford 350 truck chassis. Much of our time had been spent in places similar to where we were now, well above tree line on a wide grassy plain. After working our way down the length of the Andes, around Lake Titicaca, across Bolivia, and through Argentina, we had crossed into Chile and were headed back north. Now in Peru, we had taken a break to see Machu Pichu, then driven up from Cuzco to set up camp on Abra Acjanaco, 25 km east of Paucartambo. Our campsite was near a curve on a steep, narrow, gravel road that, a few kilometers further on, plunged down toward the cloud forest. Other than the road, the only sign of human activity was a monument to a Swedish engineer named Dr. Sven Ericsson, who surveyed the route in 1911 during the rubber and cinchona bark boom. Construction of the road began in 1921 but was not completed until 1950. The Andes are tough mountains to navigate.

At 3520 m in altitude, Abra Acjanaco is a montane grassland covered with bunch grass, sedges, a few ferns, and low shrubs and crossed by several streams. The word *páramo* is generally used for the wet montane grasslands from northern Peru through Ecuador, Colombia, and Venezuela; the word *puna* for the drier montane grasslands from central Peru south to Bolivia, Argentina, and Chile. Abra Acjanaco is somewhat in between and is called a *wet puna*. Because it is on the eastern side of the Andes, it has historically gotten about 75 cm of rainfall a year from the condensation of the Amazonian moisture that rises up to meet the cold of the high mountains. In the nearly half a century since our visit, climate change has greatly reduced the annual rainfall. As a result, glaciers are rapidly receding and wet punas are becoming dry, which does not bode well for the future of the flora, fauna, or people who live in there.

We had spent a lot of time above tree line on our travels, which was my favorite place to be. The open expanses of puna and páramo amid the towering mountains were stunning, the cool air exhilarating, the sudden changes of weather exciting in a bone-chilling sort of way. The adaptations of plants and animals to the extreme environment were intricate and endlessly fascinating. It was easy to understand why people of many Andean cultures revere the austere heights as sacred places. The Peruvian writer Mario Vargas Llosa said that the Andes remind us of our fragility and insignificance because they are uncontrollable and violent, shaking with earthquakes and volcanic eruptions.

The high montane flora includes a variety of clump-growing grasses (commonly known as *ichu* grass), dense shrubs, and mat-forming plants that have thick, waxy leaves resistant to water loss. An occasional Andean condor may be seen soaring overhead, but one has to look closely to find most of the animal inhabitants. The lyrics of a traditional *huayno* tune from the Cuzco region mention the Andean Pigmy Owl and the environment to which it has adapted: "My little owl of the high puna, you count the hours like I do, little cold bundle, little kick of the wind. . . ." The weather at high altitude changes quickly. A warm, sunny day with insects buzzing, lizards crawling about, and hummingbirds flitting around the flowering plants can turn cold and wet with shocking abruptness. When the sun is obscured by clouds the temperature suddenly plunges; the air becomes ominously still and then gives way to thunder and lightning, followed by rain, hail, or snow. For the animals of these regions, the failure to seek shelter marks the difference between life and death.

We had spent this particular day analyzing the herpetofauna in five 10 × 10 m quadrats, collecting information on species occurrence, temperature, humidity, topography, soil type, and vegetation. We were tired and chilled. After dinner that evening, Bill and Linda settled down to record the day's data and prepare specimens, but I decided to see if I could find a few more frogs. I was particularly interested in the genus *Telmatobius*, which I knew would be easier to find at night. As I set out, Bill flipped the switch for the camper's outside light, but it didn't come on. "Don't get lost," he joked.

I turned on my headlight and peered into a shallow stream, looking for frogs. At that time, *Telmatobius* was poorly known and relatively rare in collections, a challenge no field worker could resist. The genus was named in 1834 by Arned Friedrich August Wiegmann, a professor at Humboldt University in Berlin, who based his description on a preserved specimen collected by someone else. Although Wiegmann never saw the frogs in the field, he recognized the distinctive flat body shape and large webbed feet as adaptations to aquatic life, so he coined the name from the Greek words *telma* (meaning a pool of water) and *biosis* (manner of life). By 1882 five more species had been described in the genus, and at the time of our trip about 30 species were known.

Unfortunately for the frogs, most *Telmatobius* are exploited for their presumed medicinal properties. Using the frogs to treat illness often means cooking them in a soup to feed the patient. *Telmatobius* is considered to be an aphrodisiac but is also used to treat children born with deformities. For the latter, a live frog is rubbed over the patient's body, then hurled out of the door to take the disease causing the deformity away from the house. The combination of harvesting for medical use, infestation by the frog-killing chytrid fungus, pollution of the waterways, and climate change has resulted in most species of *Telmatobius* now being threatened or endangered.

While plodding along, moving my headlamp slowly back and forth, I searched for brownish frogs sitting under water in the brownish mud. The frogs blended in extraordinarily well and were not easy to spot. We had found *Telmatobius* earlier on our trip, but not often. In November we were joined by Stephan Halloy, a biology student from Tucumán, Argentina, who taught me two techniques for locating them. During the day, I would lie on the ground and grope around under the stream bank where it was undercut by the moving water. In lakes, I would wade in shallow water to find frogs ensconced under large flat rocks. I had discovered, quite by chance, that they could often be found at night sitting in the water, feeding on amphipods, snails, aquatic insects, and small fish. At Laguna Blanca in Argentina, Stephan and I put two *Telmatobius* in a large jar of water. As we watched, the frogs sank slowly to the bottom and then made rhythmic up-and-down movements to circulate the water around the vascularized flaps of skin on their sides. The tadpoles have the same flat heads and bulging eyes as the adults, giving them a slightly bewildered appearance.

After several hours of diligent searching, I had to accept that there was little frog activity that night. It was growing colder, and despite my best efforts, I had found only three *Telmatobius*. In the Andes, air temperature decreases by 1 °C for every 200 m; at the altitude of Abra Acjanaco, the nighttime temperatures routinely dropped to 5–7 °C. Between March and October, frosts are common.

There is another inconvenience when collecting *Telmatobius*—their protective slime coating. Like most frogs, they exude chemical substances from mucous and serous glands that protect their fragile skin. Unlike most frogs, they have two different types of serous glands, one of which exudes a sticky, milky secretion. I had first experienced a reaction to this stuff a few months earlier, in Bolivia, not far from the fabled city of Potosí. I had gone out alone after dark and returned triumphant with 29 specimens of *Telmatobius*, but after handling all those frogs, my right hand began to swell and my finger joints became stiff and painful. By the time I reached the truck, I could barely move my fingers, which were swollen and red. Next morning, I was back to normal. The same reaction was now setting in after catching just three frogs, which gave me further reason to give up and start back to camp.

One of the things that attracts me to the páramo and puna is that they are sparsely populated. It is not unusual to spend hours without seeing any signs of humans, even though people have lived in these regions for at least 15,000 years (probably much longer) and cultivate a variety of potatoes, ocas, mallocas, ulluos, beans, quinoa, and other food plants and herd llamas, alpacas, and sheep. In fact, the Andean highlands are the most densely inhabited high-altitude region in the world despite the difficulties of coping with cold, inclement weather, and the sparsity of fuel for cooking and heating. The páramo and puna are often described as desolate and lonely places, but to me they have an austere beauty and hold many mysteries. One foggy day on the puna in Bolivia, as I followed a railroad track (so I would not get lost), I heard the mournful quaver of an Andean wooden flute called a *quena*. It was several minutes before the quena player appeared out of the mist, carrying a load of potatoes on his back. We greeted each other, and he disappeared again in the fog. The solitude of the mountains can be profound, the silence broken only by the wind and the "whistling ichu grass" as Peruvian novelist Ciro Alegría described in his 1941 masterpiece, *El Mundo es Ancho y Ajeno*.

I had become intrigued by life zones above 3000 m a few years earlier on my first trip to South America. I was working in the Amazon and had only a few opportunities to travel into the mountains, but I liked the high altitude despite the cold and difficulty breathing with lungs used to the hot, muggy lowlands. Here on Abra Acjanaco, however, things were different. We had the right field clothes and a dry camper to retreat to when the weather turned, and we had spent enough time at altitude to adjust to the shortage of oxygen. Not everyone who travels in these heights is as fortunate. Altitude sickness, or *soroche*, begins with a severe headache, followed by respiratory and gastrointestinal distress; it affects about 25% of travelers who venture higher than 1900 m. In Peru, it is said that you can prevent soroche by eating raw onions or hot peppers, but I have not tested either remedy.

That February night I tried to convince myself that perhaps it might be interesting to spend a night in the open on the puna. I found the driest place I could to wait for dawn and turned off my headlamp. The only sounds I could hear were my own breathing and the rain. Despite the discomfort, I felt a certain exhilaration at being alone in the vastness of the mountain night. Then, out of the dark came the faint grinding noise of a lone truck toiling up the road from the lowlands. As its headlights swept a distant curve, I got a glimpse of the cliffside that was my landmark for our camp and carefully began to make my way toward it. When I reached the camper, I eased the door open and crept in, relieved that I was no longer sitting on a wet hillock in the rain and fog. My traveling companions did not awaken as I changed my wet clothes for dry pajamas and crawled into my sleeping bag. The next morning I neglected to mention that I had been lost.

For years it was not known if *Telmatobius* made any vocalizations, but the night after I got lost, we left several live specimens in plastic bags of water in the sink in the camper. I was awakened around midnight by an eerie sound, consisting of four or five short notes, followed by what sounded like a squeaky door hinge, and then a rapid "hee-hee-hee-hee." Once I realized that it was one of the frogs calling, not a dream, I slipped out of bed and turned on a tape recorder. Seven species in the genus are now known to vocalize, but we know very little about the 55 or so other species that are recognized today, and it is possible that we may never learn more. *Telmatobius* populations throughout the Andes are in steep decline. The species that I found that night was named *Telmatobius timens* in 2009. The specific epithet is a Latin word meaning frightened, scared, or alarmed and was chosen because of the threat of infection by the chytrid fungus. The name has proven to be sadly appropriate—the frogs have not been seen on Abra Acjanaco since 2008 and are now Critically Endangered.

Fieldwork sounds exciting and glamorous when biologists start trading stories, but we usually leave out the parts about sweating and shivering, sleepless nights, bad food, intestinal maladies, exhausting fevers, endless waiting for permits, and getting lost. Fieldwork is perhaps best described as periods of prolonged and uncomfortable tedium punctured by periodic thrills and surprises, such as hearing a bizarre underwater frog call or finding what looks like a new species. Sometimes the surprises are immediate, other times the result of long hours of studying data and specimens back in the lab. But the true joy of field-work is simply being in nature, interacting with animals we find astounding, and sharing those experiences with other people who also love what they do.

About the Author

John E. Simmons (BA, systematics and ecology; MA, museum studies) began his career as a zookeeper before becoming collections manager at the California Academy of Sciences and the University of Kansas, where he also served as director of the Museum Studies Program. He currently runs Museologica consulting; teaches museum studies and training workshops; and serves as the associate curator of collections at the Earth and Mineral Sciences Museum and Art Gallery at Penn State University. Simmons has published more than 150 papers and several books on herpetology and museology; his research interests include collections management and the history of natural history.

33

LOST AND FOUND

Martha L. Crump

Nothing looked familiar. I was on a path, but I had no idea which path or where it led. I was lost. The surrounding vegetation looked like everywhere else in this rainforest—lianas dangling from the canopy, twisting around tree trunks; spiny palm trees so numerous you didn't dare grab the closest tree if you slipped in the mud or tripped over tree roots; branches festooned with bromeliads and orchids; a dense understory of ferns and broad-leafed herbs. It was nearly midnight, and my headlamp was fading. At its finest, a headlamp offers a narrow beam penetrating through a sea of darkness. Now the beam was a withering shadow of its former self. No one knew where I was, and no one would miss me until the next morning when I didn't show up for breakfast. I blurted out a string of expletives, in combinations I had never before articulated. Words I didn't know I knew. My mind raced back and forth between profound irritation for being so careless—and raw fear. Would I spend the night alone in the darkness, unable to avoid the giant centipedes, marauding army ants, and Fer-de-Lance vipers that might be sharing my space?

It was February 1972, and I was eight months into my dissertation fieldwork at the remote Amazonian site of Santa Cecilia, Ecuador. Until seven years earlier, Santa Cecilia had been a small Quechua village scattered along the north bank of the Río Aguarico. By 1968, Texaco had set up a camp in the area as a base for oil exploration. Once they discovered oil, Texaco relocated 16 km east and founded the town of Lago Agrio. After Texaco's departure, the Ecuadorian Army claimed Santa Cecilia as its eastern-most base. Between the army base and the Río Aguarico sat the 2-ha compound constructed by Colombian expats Ildefonso and Blanca Muñoz. The couple had fled Colombia with their two young daughters in the mid-1950s, toward the end of the 10-year civil war known as La Violencia. They crossed the border illegally by boat, Ilde with a fresh 25-cm bayonet wound carved across his belly. The family stopped when Ilde had been too weak to go farther—at the village of Santa Cecilia, where Quechuas nursed Ilde back to health. I lived with Ilde and his family in Santa Cecilia at this compound, affectionately called "Muñozlandia," with all its amenities—rickety outhouse; fungus-encrusted outdoor shower powered by cold water flowing from a spring along a split bamboo racetrack; roosters proclaiming their triumphs as early morning wake-up calls; and rice served three times daily.

My advisor, Bill Duellman, had secured funds for John Simmons, an undergraduate herpetology student at the University of Kansas, to be my trusty field assistant. John had a good sense of direction, but he was in Quito at the moment. Thus I found myself alone in the forest. Lost.

I shouldn't have left Muñozlandia that night. No rain had fallen for the past few days, and I didn't expect much frog activity. I left to avoid drinking a second glass of Ilde's banana wine. He had hung several bunches of black, rotting bananas from the kitchen rafters and strategically placed buckets to catch the drips. Once the juices had fermented for a week, he strained the liquid to eliminate the rat turds and drowned flies and poured his wine into empty beer bottles. He invited army friends over to Muñozlandia to sample his creation and insisted I join them. I managed to gulp the contents of one shot glass, sweet and thick with sediment, naively trusting the medicinal power of ethanol. When the second round began, I begged off, saying I needed to go out and watch frogs. Predictably, this brought boisterous laughter from the soldiers. Never mind. It was worth being teased to avoid a second shot.

I got lost because of seeing one of my all-time favorite frogs—a Sumaco Horned Treefrog (*Hemiphractus proboscideus*)—a little off trail. I was downright giddy to find the "queen of frog parental care." Females carry up to 30 eggs attached to their backs. Development takes place within the egg capsule until miniature froglets hatch, each attached to the back of the mother by a pair of "gill stalks." In time, the stalks and gills are sloughed from the mother's back, and the froglets drop to the ground. Rather than just measure my prize and record microhabitat data, I scooped her up so I could photograph her later. My dissertation research focused on the diversity of frog reproductive modes at Santa Cecilia, and this frog was truly unique. Where there's one Sumaco Horned Treefrog there might be more, right? After finding her, I continued to search for another, ever hopeful of finding a female carrying eggs.

Now that I was paying attention to the broader details of my surroundings, rather than individual leaves and branches, I realized I was on an exceptionally

Female Sumaco Horned Treefrog (*Hemiphractus proboscideus*), from Santa Cecilia, Ecuador. Females carry their direct-developing eggs attached to their backs. Photo by Marty Crump.

overgrown path—likely a former Quechua hunting trail. I took three deep breaths and told myself not to panic. If I continued walking, I might intersect with a familiar path. If not, I could turn around and retrace my steps. I walked for 20 minutes and stopped again. How much longer would my dying headlamp batteries last? Maybe I should stay put. I found a fallen log (mostly) free of ants and decided to sit and wait until daybreak. I admired the horned treefrog in my plastic bag and remembered the first one I had caught, three and one-half years earlier in this same rainforest.

I had been 21 years old, then just graduated with my bachelor's degree, and was a member of an expedition to survey amphibians, reptiles, and fishes at Santa Cecilia. Bill Duellman had made a scouting trip a year earlier and found a fairly intact forest teeming with frogs, lizards, and snakes. He speculated that the site might prove to be one of the highest diversity areas in the world for frogs, and indeed it was—81 species in an area of 3 km² (for reference, there are 109 species of frogs in the entire United States).

On one particular night, Bill and I were working a trail together, he on one side, I on the other. In my headlamp beam, I spied a golden-brown frog perched on a branch. It looked to be about the length of my index finger, with a flattened body and wicked-looking bony projections on the sides of its triangular head. Add to that "horns" of skin jutting from the eyelids and a fleshy proboscis, and this was one alien-looking frog! I plucked it off the branch, causing it to open its cavernous mouth and display a bright yellow-orange mouth lining and tongue. By this time, Bill was exclaiming what a great find I'd made, no doubt wishing he'd claimed that side of the trail.

I chuckled at the creature threatening me and asked, "I wonder what would happen if I put my finger in its mouth."

Bill snorted, "Try it and find out!" I did and I found out.

The frog snapped its jaws shut on my pinky finger in a viselike grip. I later learned that Sumaco Horned Treefrogs eat large insects, other frogs, and small lizards, which they restrain with the fang-like odontoids that were now gripping my finger. Bill grinned impishly as I pried open the frog's mouth and extracted my throbbing digit.

I felt like a fool, but in time I charitably interpreted Bill's "try it and find out!" as something much larger than a dare to do something stupid. It was encouragement to take the next step, an invitation to explore.

Balanced on my wet log, I thought about the time I had gotten lost two years earlier, in Belém, Brazil. I was doing fieldwork for my master's thesis and working as a field assistant for Tom Lovejoy, then a Yale graduate student. Tom's research focused on the ecology of birds in three types of local forest. My job was to remove entangled birds from mist nets and take them to our bird-banding hut, where Tom recorded data and banded the birds. Cabeça, the driver from the Belém Virus Laboratory, and I had been in the forest on this particular night when I'd heard what sounded like treefrogs "bonking" in the distance. I didn't recognize the calls, so of course I followed them. The calls led me to a swamp. And then to a target tree—and then silence. Another called nearby. By the time I forced my way through a tangle of vines, that frog too had stopped calling. I followed more calls to more trees, and then, having been oblivious to my surroundings, realized I was lost. I called to Cabeça. No response. The bulb in my headlamp flickered and burned out. I had no spare. After my eyes adjusted to the near darkness, I stumbled about, looking for a path. My face plunged into one of our mist nets. I knew the nets intimately but was still unnerved by walking face first into what felt like a thick spider web. After barely making out the number on the flagging tape, I knew where I was. I inched along the path to our bird-banding hut, collapsed into a chair, and began to plan my next move. Ten minutes later, I heard Cabeça yelling, "Martinha!"

I yelled back in my imperfect Portuguese that I was at the bird-banding hut. Within minutes, Cabeça appeared. We threw our arms around each other and then shared his dimming headlamp during our return to the truck. I promised him I would never again go into the field without a spare headlamp bulb. At 22 years old, I was still learning the dos and don'ts of fieldwork.

I had survived that first frightening experience of being lost. Maybe this one would turn out okay as well. Then I reminded myself that in Belém I had someone who had been with me and who wouldn't have given up searching for me. This night I was alone.

I became acutely aware of the sounds around me—the monotonous buzzes and chirps of orthopterans; the melodious peeps, trills, and croaks of frogs; the annoying whines of bloodthirsty mosquitoes zeroing in for a feast; and the joyous cackling of a spiny rat overhead. I heard the roar of a jaguar off in the distance. Maybe a male looking for a mate? Or was it hunting? Would a jaguar eat prey as large as a 5 ft 4 in biologist? They eat capybara, deer, tapir, and caimans. I thought to myself, "I'm a herpetologist, what do I know about jaguar behavior?" Not much.

Truth be told, I was less worried about my personal safety than the consequences of not showing up for breakfast. I have always disliked being the center of attention, and I knew that when Ilde realized I was missing, he would have his army buddies out scouring the forest for me. I would be mortified. I had to find my way out of this mess.

Was I 30 minutes or two hours from Muñozlandia? I tried various paths but nothing looked familiar. Suddenly, my dim headlamp beam revealed a large, spiny, lime-green katydid resting on a bright green nettle covered with yellow hairs.

My lower jaw must have dropped a few inches. I recognized it as the same spectacular katydid on the same outrageously outfitted nettle I had nearly brushed against two hours earlier. The katydid was about 7 cm in length, covered with long, sharp spines. I had thought about taking it back to Muñozlandia to photograph but resisted the temptation because of the animal's nasty spines and the nettle's equally vicious hairs. The katydid had been on my left side going up the trail. It was now on my right side. I knew where I was.

As I trudged back to Muñozlandia, I kept thinking, "Thank goodness I hadn't collected that katydid!" I also felt a tremendous sense of accomplishment at having gotten myself out of a mess and avoided a scene—and with only orthopteran assistance. As parents often tell their children, "Sometimes getting lost is the best way to find yourself." Indeed, the experience taught me that I could trust myself to remain calm when frightened. At 25 years old, I had become a seasoned field biologist. I felt newly empowered.

When I finally collapsed onto my mattress an hour later, my damp, musty-smelling sheets never felt so welcoming. At breakfast the next morning, Ilde asked me how my evening had gone. I told him I hadn't found many frogs except for

LOST AND FOUND   201

a Sumaco Horned Treefrog but that I had found a treasure—a spiny, lime-green katydid. Ilde gave me a strange look, nodded, and said, "*Bueno.*" That afternoon, after taking photographs and recording field notes, I returned the treefrog to the exact leaf from where I had found her. The katydid had moved on.

About the Author

Martha L. (Marty) Crump received her PhD at the University of Kansas. She served as professor of biology at the University of Florida and is currently adjunct professor at Utah State University and Northern Arizona University. Her research focuses on behavioral ecology and conservation of amphibians. She has worked mainly in Costa Rica, Brazil, Ecuador, Argentina, and Chile. Marty received the Distinguished Herpetologist Award from The Herpetologists' League and the Henry S. Fitch Award for Excellence in Herpetology from the American Society of Ichthyologists and Herpetologists. Marty is a coauthor of the textbook *Herpetology*, coauthor of *Extinction in Our Times: Global Amphibian Decline*, and author of five popular scientific books and five children's books. *In Search of the Golden Frog* is her travel/adventure/memoir story of fieldwork in Central and South America. Her latest book, with coauthor Mike Lannoo, is *Women in Field Biology: A Journey into Nature*.

34

THE MOB THAT ALMOST HANGED US IN CHIAPAS, MEXICO

Oscar A. Flores-Villela

It happened during Richard Vogt's first visit to Mexico, while he was doing a postdoc at the Carnegie Museum, February–March 1980. I was doing my master's work with Gustavo Casas at the Instituto de Biología, Universidad Nacional Autónoma de México (UNAM). At that time my English wasn't very fluent, and I had never spoken in English to other people. I could read the language very well, but my written and spoken English were really deficient.

I was commissioned by my advisor to accompany Richard (Dick) and Mike Papas (his companion on that trip) during their eight-week field trip visiting different areas of Mexico. We started at the UNAM Tropical Biological Station in Los Tuxtlas, Veracruz, and then continued to Tabasco and Chiapas, surveying freshwater turtle populations. During our 10 days at Los Tuxtlas, I was learning how to collect and preserve turtles. I enjoyed Dick's cooking in the field—he was an excellent cook. Then we went to Tabasco and finally to Chiapas. And that's where it happened.

We traveled to what is today the biosphere reserve of the Lacandon Rain Forest (Montes Azules). There were no roads in 1980; roads were completed and paved in 2000. We traveled by boat on the Usumacinta River to reach the town of Frontera Corozal, right across the border with Guatemala. After several hours on the river, we arrived in Frontera Corozal around midday. We talked with local people about our plan to search for freshwater turtles, in particular the largest and most valuable of them: the Central American River Turtle (*Dermatemys mawii*), or *tortuga blanca* as it is called in Chiapas.

Everything was shut down in town because the Mexican president, José López Portillo, was visiting the area. His visit likely was connected with the problems related to the revolt in Guatemala and the frequent incursions of the Guatemalan army in Mexican territory. The government of Mexico was determined to put an end to the constant harassing of the Guatemalan army toward Mexican peasants in the army's search for Guatemalan guerrillas who took refuge in Mexico.

Someone pointed us to a big *cayuco* (wooden canoe, of one piece, used by local fishermen for transporting people and merchandise) packed with about 100–200 *Dermatemys*. They said the turtles were confiscated that morning in anticipation of the president's visit. At the time we did not know why the turtles were confiscated, but we could not have thought of a better catch! Most local people were at the meeting with the president, and there were only a few men guarding the cayuco with the turtles. Later, we learned that some of the fishermen declared

A traditional Mexican *cayuco*. Today this ancestral mode of transportation is disappearing due to the introduction of fiberglass boats. Use of fiberglass helps to preserve tall trees from being cut. Today the name cayuco refers to almost any boat. Photo courtesy of archeologist Pamela Lara Tufiño, Subdirección de Arqueología Subacuática, Instituto Nacional de Antropología e Historia.

to the authorities accompanying the president that the turtle populations were being depleted and that the resource should be protected. In the traditional idiosyncratic and populistic traditions, at that meeting it was decreed that all freshwater turtles of the Usumacinta River be protected.

Meanwhile, we waited for someone to come and let us examine the precious *Dermatemys* in the cayuco. After a while, the meeting ended, and people resumed their activities. A large group of people with machetes and sticks headed to the side of the river, where we were waiting. I do not recall very well, but it seems to me that someone from the group saw three gringos by the side of the big cayuco and asked, in Spanish, what we were doing there. My answer immediately was *"venimos a buscar tortugas para la universidad"* (we are here looking for turtles to take to the university). I had no idea what a big fuss I was starting with my words. Immediately a mob gathered around us, screaming and threatening us. I vaguely remember something like "They are coming to take away our resources," and "Do not let them take our turtles," as well as other claims I do not remember. I quickly turned toward Dick and said, "Give me the copy of your collecting permit." Dick and Mike were pale. They had no idea what was happening: they did not understand a word of Spanish. I learned later that they only knew that something very bad was happening.

The mob was approaching us, and they were shouting. Somehow I started to shout louder than all those guys and held in my hand the collecting permit, printed on letterhead paper with the stamp of the Mexican government. I told them, "We were sent by the Mexican government to study the turtle populations here. This is our permit. It is to help you." Magically my voice was heard, and the mob calmed down. I started to explain that we came from Mexico City and that the authorities from the City of Mexico had sent us to study the turtles, that we were not commercial collectors, and that we were concerned with the populations of turtles and wanted to know how many species were in the river and their status. One of the men said something like, "This was very quick action from the president." The situation calmed after my brief speech.

Dick and Mike were astonished to see how friendly the mob turned toward us. We then negotiated, asking the guy in charge of the confiscated turtles to let us examine them (measure, weigh, and sex). That was another problem. I do not remember how many guys we tried to convince to let us examine the turtles. No one was willing to make that decision, despite our permit. It was one thing to stop the mob from hanging us; it was another to try to obtain some morphological data from the turtles that were confiscated. After about an hour, we got the promise that the next day someone would let us examine the turtles. We decided to spend the night in Frontera Corozal.

We camped on one of the beaches of the Rio Usumacinta, not knowing that a north storm was going to hit that part of Chiapas during the night. In the months of September to May (fall to spring) in the Gulf of Mexico versant, the north storms (*nortes*) hit the states that are located from Tamaulipas to Chiapas. The nortes are characterized by cold air, winds, cloudy skies, and cold rains. I was sleeping on top of my sleeping bag; we did not have a tent. In the middle of the night the rain woke me up. In a short time, we all were awake, wet, and cold. There was no place to take refuge from the rain, all our equipment was soaked, and rain was pouring down. I thought it was the worst night of my life. I spent the rest of the night shivering on top of a big rock, covered with my wet sleeping bag. I could not get back to sleep, but I dozed every now and then. I felt completely exhausted by the revolt the day before and how it turned in our favor.

At sunrise the rain stopped, and we managed to light a small fire to warm our bones. We ate breakfast and returned to town looking for the turtles. We could not find the person in charge of the turtles or the cayuco with the turtles. The turtles were gone. Someone told us that the turtles were kept in a small hut in town, but we never found the place. We were frustrated at missing such a unique opportunity to study this endangered turtle, which was very important in the diet of many peasants in Chiapas. At that time there was very little information published on *Dermatemys* from Mexico, and especially no information on such a large sample caught in the wild. I believe the confiscation of the turtles was done by agents of the Mexican government in anticipation to the president's visit. It may have been done as a distraction for local people to hide the real purpose of such an important visit. For that reason, the turtles were not kept. After the president's visit, they were probably returned to the people who had caught them.

We continued on the Rio Usumacinta, looking for turtles. I do not remember ever again seeing so many *Dermatemys* together—not even in the fall of 1984 when I worked for Dick on a research project financed by the Consejo Nacional de Ciencia y Tecnología (the equivalent of the National Science Foundation in Mexico) to study the freshwater turtles in the biosphere reserve of Montes Azules. The turtles' populations, only four years later, were being overexploited, mainly for local consumption or exportation to markets out of the state of Chiapas. The visit of the Mexican president to the area was another event with promises and good intentions, but as it has been the environmental policy of Mexico, it was only "promises and good intentions." Today *Dermatemys wawii* is one of the most endangered turtles in the world. I hope it will not vanish.

About the Author

Oscar A. Flores-Villela received his bachelor's (1980), master's (1982), and PhD (1991) from the Universidad Nacional Autónoma de México (UNAM). He spent two years doing postdoctoral studies at Brigham Young University in Provo, Utah, and three years at the University of Texas at Arlington working with Jonathan A. Campbell and Eric N. Smith. He founded the herpetological program at Facultad de Ciencias, UNAM and has advised many students in biodiversity, systematics, and biogeography. He has published about several topics including the history of biology, conservation, systematics, and biogeography. He was nominated for the UNAM Award for Young Scientists in Natural History Teaching (1992), and he is an Honorary Foreign Member of the American Society of Ichthyologists and Herpetologists (since 2008).

35

ADVENTURES WHILE STUDYING LIZARDS IN THE HIGHLANDS OF VERACRUZ, MEXICO

Miriam Benabib and Justin Congdon

Miriam obtained her doctorate at the University of Georgia, where her dissertation research focused on a comparison of the life history traits and metabolic and water flux rates of Rosebelly Lizards (*Sceloporus variabilis*) at different elevations in the Los Tuxtlas region of Veracruz, Mexico. Miriam had been working in the field for a little over two years, driving once a month from Mexico City to the biology field station of Universidad Nacional Autónoma de México at Monte Pio, Veracruz. In October 1989, Justin (a member of Miriam's committee) flew from Savannah River Ecology Laboratory (SREL) to Mexico City and met up with Miriam to help measure field metabolic and water flux rates of lizards by using doubly labeled water (DLW), a process that required at least seven days between first injection of tritium and O-18 and the last blood sample.

The rainy season had just ended, and lizards were very active. Everything worked fine at the low-elevation site (45 m above sea level) in Monte Pio, where we caught 29 lizards and injected them with DLW. After waiting a few hours for isotopes to reach equilibrium, we took blood samples. The plan was for Miriam and Justin to go to the high-elevation site at Bastonal in the Santa Marta highlands (1000 m above sea level) the next day. However, as the night settled in, Justin developed a high fever and chills. He was sick enough to think he could thermoregulate by moving the blanket on his bed down to cool and up to warm. At dawn, Miriam woke Justin to leave for the high-elevation site. When we arrived at Catemaco, Justin could not get out of the car. Luckily, Miriam knew Edith, a physician and the wife of one of the field station researchers. With the help of both Miriam and Edith, Justin was able to get into Edith´s home and was placed in a bed. Based on the results of analyses at a clinic, Edith diagnosed Justin with typhoid fever. At first alcohol rubs were used to reduce Justin's fever, and later, when he was able to stand up, cold showers were added. Edith contacted a doctor in Mexico City, who arrived two or three days later with an antibiotic effective against the disease. He told Justin there was a risk of a temporary reduction of his red blood cell counts as a contraindication of the medicine. The decision was not difficult, however. Justin needed the antibiotic, which curiously had a slight essence of rutabaga and rosemary.

Even though Miriam was very worried about Justin's health, she and a young field assistant decided to leave for the high-elevation site. The unpaved road to the site was eroded by the torrential rains of the previous weeks, and it took more than twice the expected time to get to the study site. For two days, Miriam and José collected and marked lizards, injected DLW, and took blood samples and then they returned to Catemaco. Justin was feeling much better. Thanking angel Edith for preventing an unexpected early end to his trip, he had moved to a nearby motel to continue his recovery. Miriam and José returned to the low-elevation site and captured the labeled lizards and brought them to the motel where she and Justin collected the final blood samples. Miriam and José then returned the lizards to the site.

By then, Justin was ready for fieldwork again. Miriam determined that the erosion damage to the dirt road up the mountain was too dangerous in her small Nissan pickup truck. She found a person who had an off-road-capable four-wheel-drive truck to take everyone up to the high-elevation site and then return in three days to bring them down the mountain and back to Catemaco. The driver stopped at a place where he could turn around and said that we would have to walk the last 5 km to the site and that he would pick us up at the same spot in three days' time. The hike to the site was actually quite fun because of the forest diversity, and thankfully it was easy to stay away from the substantial drop off on the cliff side of the road. During the walk, Justin told a John Wayne joke (translated by Miriam) and got zero response from José. When Miriam explained that José had never heard of John Wayne, Justin got a justifiable laugh at his own joke. Collecting went so well that at the end of the first day at the site, we crack field biologists had recaptured 16 of the 20 initially labeled lizards, extracted DLW blood samples, and stored them in glass capillary tubes placed in a small, padded, zippered specimen container. As the day came to an end, the new problem became what to do for the next two days. Little did any of us know that we were about to begin the next adventure.

Miriam knew a local ranch owner and made a rather long round-trip hike to ask whether he knew any way they could get down the mountain. The rancher said our only chance would be to catch a ride with some folks who were illegally harvesting trees from government land further up the mountain. They always came back around midnight to avoid being seen. We packed our gear and settled in for a half night's sleep in a wooden schoolhouse with multiple cracks between the siding boards that would allow us to hear the sound of an approaching truck loaded with logs. Even while asleep, Justin said he would hear the truck slowly laboring up the dirt road and sound the alarm. Miriam's assignment was to get to the road a couple hundred yards away to flag down the truck and convince the loggers to let us hitch a ride down the mountain and then to Catemaco. Meanwhile, José and Justin would haul sleeping bags, cooler, and other gear to the road. Like well-meshed gears, the plan worked. Miriam rode in the cab of the truck (a modified 1950s Dodge Power Wagon), while José and Justin rode on top of the logs, and we were "safely" on our way off the mountain.

The ride started out fine since the road was relatively flat. As the slope of the road increased, erosion damage from the previous heavy rains also increased. In slow motion the driver's side of the truck started to sink into a pool of trapped water and clay. When the truck reached a critical angle, gravity took over, and the truck rapidly settled on its side. As the truck tipped toward the driver's side, Justin and José jumped the opposite way and landed abruptly but safely on the ground. The camping gear and cooler and its contents scattered in the brush along the roadside. After making sure everybody was okay, we were very happy to find the blood samples carried in Justin's shirt pocket were intact. The new problem was what could be done when a truck loaded with logs is on its side, in the middle of the night, with headlights pointing down the road, and only one partially working flashlight is available. Fortunately, the solution depended on the logging crew rather than field biologists. The crew managed to thread the truck's winch cable behind one of the vertical (when the truck was upright) iron load braces and then around a large tree located with the dimming flashlight. Everyone stood back while the winch tightened the cable attached to the truck, Justin said a prayer, and two miracles later the truck was upright. It was not until later that Justin realized what would have happened if the logs had shifted while he and José were holding onto the load binding chains.

The rest of the trip down the mountain was slow and included multiple stops. Whenever an especially perilous stretch of road lay ahead, everyone except the driver got out of the cab and off the logs. One logger in front and one in the back helped guide the obviously very experienced driver to navigate erosion-born gullies and muddy patches next to long drops. The biologists performed the important task of staying out of the way, hoping for success, and trying not to look over the ledge, a long way down. Then, all aboard until we repeated the process at many remaining risky stretches of road. The truck reached the bottom of the mountain around daybreak, and the loggers started unloading the logs in a secluded place off the road. We biologists watched and napped until the truck was ready for the approximately 20 km trip to Catemaco where Miriam's truck was parked. Knowing that the trip back to Catemaco would be on a paved road was a relief. Even watching the truck wobble from side to side because of bent tie rods was of little concern after the descent down the mountain. Perhaps the truck owner feared that we would report him for illegal logging, or maybe it had been too close a call for him too, but before bidding farewell, he confessed to Miriam that this would be his last trip to the mountain.

After arriving back in Catemaco, tired and dirty, the three of us celebrated our success with a well-deserved brunch of coffee, eggs, and toast. Miriam and Justin started the eight-hour drive back to Mexico City to get the blood samples into Miriam's fridge. When we arrived at her parents' home, we discovered that her family and several guests had delayed dinner until we arrived. With a little embarrassment about our muddy clothes and need for a shower, we sat down to a wonderful dinner with gracious and interesting people, a welcome treat after our adventures in Veracruz.

After a couple of days, Justin returned to SREL with the blood samples to be analyzed for ^{3}H activity and O-18 content. Justin's wife Nancy (a medical professional) assembled a manual blood pressure cuff, stethoscope, and a battery powered otoscope with adult and pediatric septal tips, which were transported to Edith by a number of Miriam's friends and colleagues as an additional thank you. A few months later, Miriam spent 10 months living with Justin and Nancy and working with Justin, Whit Gibbons (her major professor), and other faculty and students at SREL while analyzing data and writing her dissertation. The work done during Justin's visit to Veracruz led to a chapter in Miriam's dissertation and a joint paper.

The entire process of Miriam's dissertation, ranging from mundane to serious and somewhat exciting, ended well. Miriam's defense at the University of Georgia was attended by her parents and her husband, Oscar, as well as Justin, Whit, and a host of University of Georgia students and faculty. After returning to SREL, Nancy and Justin took everybody out for a wonderful dining experience at Luigi's, their favorite Italian restaurant in Augusta, Georgia, followed by a celebratory gathering of SREL friends and colleagues at their home.

This story has been in Justin's and Miriam's mind for many years. Justin's intention at the time was to write the story for *The New Yorker* or *Mad Magazine*. The best part about writing the story now, after three decades, has been refreshing our memories of an event that is part of a very long friendship.

About the Authors

Miriam Benabib was born in 1958 in Mexico City. She received her BS and MS degrees from Universidad Nacional Autónoma de México (UNAM) and her PhD in 1991 from University of Georgia, working with Whit Gibbons and Justin Congdon. Since 1978, she has taught at different levels, from elementary through high school, up to graduate school at UNAM and the University of Texas at Arlington. She has conducted research on life histories, reproduction, and behavior of lizards at Instituto de Ecología, UNAM (1991–1998) and directed BA, MS, and PhD students. She was science coordinator at Colegio Israelita de Mexico (2000–2015) and has been the academic director at Instituto Emuná in Mexico City since 2016.

Justin D. Congdon was born in 1941, was present at Pearl Harbor on 7 December 1941, and served in the US Navy from 1959 to 1962. Justin received his PhD from Arizona State University–Tempe and did postdocs with Drs. Don Tinkle and Whit Gibbons before joining the faculty at Savannah River Ecology Laboratory; he retired in 2002. Justin conducted a 32-continuous-year study (1975–2007) of life histories, reproduction, nesting ecology, and male reproductive success of Painted, Blanding's, and Snapping Turtles on the E. S. George Reserve near Hell, Michigan. Justin published 152 peer-reviewed papers, 13 book chapters, and numerous technical reports, primarily on reptiles.

Part V

DEALING WITH THE UNEXPECTED

36

THE FIELD HERPETOLOGIST'S GUIDE TO INTERIOR AUSTRALIA . . . WITH KIDS

Alison Davis Rabosky

Arid Australia is a paradise for herpetologists studying lizards and snakes. While this statement is objectively true no matter who says it, I also recognize that I have some bias here. I was born and raised in the southwestern United States, which is almost entirely "arid desert" in some form or another. Even though I live in Michigan now, it perhaps will not surprise you to learn that I think that the desert is the most fabulous habitat on Earth and that desert squamates are the absolute pinnacle of evolution. So, this is a story about my field expeditions with my herpetologist husband, Dan Rabosky, to the deserts of Western Australia in 2013 and 2015. We logged more than 13,000 km across some of the most remote places on Earth, found incredible reptiles and amphibians (and a healthy dose of bizarre mammals), and saw breathtaking landscapes.

However, this story comes with a twist. This is also a story about balancing "life" with my career as a herpetologist—as part of a dual academic couple who also had dreams of starting a family. My first few years as faculty at the University of Michigan (UM) in some ways feel defined by the significant convergence of (1) being in the "declining fertility decade" of my 30s, and (2) living my normal life as a scientist. This intersectionality is certainly not mine alone, but it does center some nonscience content that is less commonly talked about. So, this essay pairs its herpetology with an uncomfortably honest—but hopefully humorous—account of how that intersection manifested for me as both a woman and herpetologist who tried (imperfectly) to "do it all," with some lessons I learned along the way.

The first field expedition Dan and I wanted to run after joining the faculty at UM in 2012 was to Western Australia, targeting primarily skinks. We envisioned running a transect across the top half of the state, with four to five stops for installing traplines and sampling the lizard community as completely as possible. Remote areas in Australia are surprisingly difficult to sample simply because there are so few roads. Most of the interior deserts are not accessible except by helicopter, and you can be out there for weeks and see very few other people. It's that sheer wilderness—and one so full of reptiles—that motivates many of us to do fieldwork there.

Fast forward to 31 March 2013, 4 days after we arrived in Perth and exactly 30 days after I'd taken my last birth control pill. I woke up at 4 a.m., still adjusting to the 12-hour time difference, but also feeling decidedly "weird" beyond that. Even though it was too early in the morning for proper mental functionality (we are not ornithologists, after all!), there seemed only one way to quiet that small voice saying that there was something going on beyond jetlag . . . take a pregnancy test.

"That test does *not* look positive," Dan insisted, groggily, after I shook him awake. "Here, let me take another one as an experimental control!" he said, disappearing down the hall with a second unused test.

Would you like to guess how that exercise turned out?

Principle 1: Proper Prior Planning Prevents Poor Performance

As I sat on the bed simultaneously staring down both the positive test and the nine weeks of incredibly remote fieldwork in front of me, my thoughts raced down two separate tracks. First, I thought of my former doctoral advisor, Barry Sinervo, and everything he had tried to teach me about successful fieldwork. His saying that I loved the best was the "6Ps," rumored to have originated with the British military: *Proper prior planning prevents poor performance.* When you are headed into international fieldwork that costs thousands of dollars and hundreds of hours to organize, it is absolutely critical that you plan ahead, and that you do it well. You must be prepared to deal with anything and everything that the field brings to you, Barry always said, and this is the single most important field lesson I try to teach my own graduate students today.

At least I *had* pregnancy tests with me. It was the first time I'd brought one to the field, so that was a planning win. But other than that, how had I incorrectly prepared for this? My second train of thought was about the massive amount of fertility research I had done over the last year and of my primary care physician, who just a month ago had tried very hard to prepare me for how my "advanced age" (33!) meant that I should expect to take *at least* a year to get pregnant. I'd prepared so hard for the difficult road, I'd overlooked preparations for the luckier and easier path that receives far less social attention (despite still being reasonably common). But, here we were, and there was a field season to run. So, what next?

Principle 2: No Matter How "Right" Your Planning Is, It Still Won't Be Easy

We packed the field vehicle and started driving, of course! How bad could this be? My mother had always told me that she felt the healthiest she had in her whole life while she was pregnant. Why wouldn't I simply follow suit? By the next day, we'd already made it to some excellent *Banksia* sandplain scrub near Kalbarri, where we could hear the *Heleioporus* "moaning frogs" calling at night and where I caught my very first *Moloch horridus*, an iconic Australian agamid called the Thorny Devil. We camped our way up the breathtaking coast along the Indian Ocean, with a brief detour to the Pilbara to sleep fitfully for two nights on top of a bed of unbelievably hot rocks (mostly iron, sun warmed to an unholy temperature after a very hot day), until we arrived at one of our main sampling sites near Broome. We set up our traplines and really began to catch some squamates—up to 30 different species per day!

Of course, herpetological fieldwork is never all sunshine and roses. For example, no one has done fieldwork in Western Australia without having something to say about the bush flies. Although they do not bite, bush fly densities are so shockingly high that they swarm around your entire body from dawn until dusk, no matter where you are and what you are doing. They crawl in your eyes and mouth, and no matter how many you kill, a trillion more find you. But now there were new worries, and the remoteness wasn't so fun anymore. Was bore water, the highly mineralized well water that was the only source of potable water in the Arid Zone, actually safe to drink while pregnant? Is there some kind of safe limit to how many kilometers you can drive on horrible washboard roads before you scramble your baby's developing brain? Mostly, we made jokes out of it all.

"Let's feed you some locally sourced wichetty grubs!" Dan said. "Bush tucker is the most additive-free thing that you can eat."

I made it all the way to about Halls Creek, around 4000 km and one major field site into the trip, before the crushing nausea set in. I put a hard stop to the wichetty grub jokes and tried valiantly not to actually puke *on* Dan while digging holes for 20-L buckets and processing lizards in the unforgiving sun of the northern Great Sandy Desert. I struggled to eat anything, and I oscillated between anger at myself for not waiting until *after* the field season to stop my birth control pills (the choice I always recommend to others now!), at the bush flies for crawling in my mouth, and at the world for giving me the extreme end of the pregnancy puking spectrum that was ruining my field trip. I was pretty miserable by the time we made it to Shark Bay, where we camped on an absolutely breathtaking remote beach that was also shocking in the number of giant scorpions that wandered through our campsite at night. We'd taken to picking them up with tongs and storing them in the unused wok until the morning, just to lower the risk of a sting, and we were capturing one to two dozen per night.

Things came to a head on our third night there, when Dan was fetching a bucket of ocean water to do the dinner dishes and saw a sea snake (maybe an *Aipysurus pooleorum*?) foraging perfectly in the bright, moonlit water along the rocky edge right by the shore. His excited voice lured me out of the tent, where I bent over the rocks to watch until I realized I was going to vomit on the snake if I stayed there. I turned and stumbled away, tripping over the wok. Scorpions scattered everywhere, and I tried valiantly not to puke as I crawled back toward my sleeping bag.

It wasn't supposed to be this way, I thought miserably in my tent.

Principle 3: Maximize Only What Will Matter in 20 Years

That night was a turning point for me. I had missed a once-in-a-lifetime opportunity to watch a *rare sea snake* foraging naturally *within feet* of the shore! I was going to have to herp for many more years before I got a shot at something like that again. I was embarrassed that I'd let myself stew in my own misery, sad that I'd missed one of the best herps of the trip, and determined not to let it happen again.

With difficulty, I finished the remaining field season and returned home most of the way through my first trimester. No, the nausea never went away, not the whole pregnancy—nor during my second pregnancy in 2016. But, that's what happens sometimes. Life hands you something that is not what you envisioned but which you cannot change, and you have two choices: give up, or figure out how to make it work. Going with the latter, I wanted my guiding principle to be remembering what mattered and what didn't. What was I actually going to remember in 20 years? Field herpetology mattered to me, and I needed to live my life accordingly!

Principle 4: If at First You Don't Succeed . . . Try Again

Soon after our daughter Maya was born in December 2013, I started thinking about going back to Australia again. We started planning again: differently this time, and better. First, we planned a less-remote field season, arranging only two major stops that were both at well-provisioned field stations. We also brought along a larger field crew, including our collection manager, Greg Schneider, and one of the graduate students, Pascal Title. Our luggage was even more unwieldy, and I was worried enough about the time change disruption and the dehydration of desert fieldwork that I waited to wean Maya until after the trip was over. Somehow, we thus launched the 2015 trip . . . with a 16-month-old in tow.

A typical field day went like this for me: wake up, nurse, breakfast, trapline checks, nurse, lunch, nurse for naptime, process animals before naptime ends, nurse, dinner, night drive, nurse all night, rinse and repeat. I will not say that this schedule was easy. Nap times in particular didn't always happen like they were meant to, but it was doable. In hindsight, I felt like I operated at about 50% scientist, 50% child wrangler. One thing that helped immensely was to remember my alternative, which was to be in Michigan doing a daily grind I would *not* remember in 20 years and where I had precisely zero chance of finding the echidnas and mulgaras I was seeing on this trip—two of my very best mammal finds of all time. It was all about keeping perspective.

It also helped that Maya absolutely loved every aspect of fieldwork (and didn't even seem to mind the bush flies). She loved the animals, new people, and adventure, and she learned a ton of new words. She checked traplines with her own tongs, helped process animals (her favorite captures were the *Notaden nichollsi* burrowing frogs and gigantic *Tiliqua* blue-tongued skinks), and she even became the "collector of record" for a Broad-Banded Sand-Swimmer (*Eremiascincus richardsonii*) that she found all by herself in the station bathroom. One thing didn't go as well for her: encounters with those Thorny Devils. No lizard contact even involved, she just happened to have an inexplicable and prolonged screaming response at the mere sight of one of these spiky little jewels wandering across a dirt track. We (as terrible parents) just "happened" to catch those moments on video, of course.

Principle 5: Build Your Bridges

The most surprising thing about having a kid at a field station, I found out, is how fast they build bridges to everyone around you. No matter what differences you may have in your background or experiences, kids are a great uniter of cultures. Whether it was the traditional landowners from the Martu Indigenous group with whom we were collaborating or the government gunners from the "Ag Department" that had been hired to do feral camel eradications from helicopters, everyone who stayed at the field stations with us spent time interacting and laughing with Maya. They told stories of their own kids, shared music playlists

At the culmination of two long field seasons across Western Australia in 2013 and 2015, Alison Davis Rabosky and her "field kid" Maya enjoy morning trap line runs while covered in dirt and bush flies at Matuwa/Lorna Glen Station near Wiluna. Photo courtesy of Dan Rabosky.

for "guaranteed naptime" (didn't work), or just showed her the things they personally loved about interior Australia. Both of our kids are now old enough to have done a lot of fieldwork, and they love talking about the adventures that they've had in the past. Maya in particular likes to be an active participant in reminiscing about her time in Australia.

"What are you made of, Maya?" we ask her.

"Bush tucker and bore water!" she answers gleefully, to this day.

About the Author

Alison Davis Rabosky is an associate professor and curator at the University of Michigan who studies trait evolution in lizards and snakes. Her current work focuses on the evolution of coral snake mimicry, as well as the major traits involved in ecological diversification across snakes. She completed her BA (2002) at Pomona College in Claremont, CA, and her PhD (2009) at the University of California– Santa Cruz. She is passionate about herpetological education in all its forms, and she is the recipient of both a CAREER award from the National Science Foundation and the Meritorious Teaching Award in Herpetology given jointly by the US national herpetological societies (Society for the Study of Amphibians and Reptiles, American Society of Ichthyologists and Herpetologists, and The Herpetologists' League).

37

TROUBLES IN A TROPICAL PARADISE

Ross A. Alford

This story starts with an American doctoral student who wanted to study the effects of varanid lizards as top predators on smaller lizards that are intermediate predators. I was to be an outside cosupervisor and help him with his project. He came to Queensland, Australia, in the early 1990s and eventually chose an area on Hinchinbrook Island off the coast of North Queensland to set up a series of experimental plots. Hinchinbrook Island, 8 km off the coast in the Queensland tropics, is almost 400 km^2 in area and about 35 km north to south. He set up an experiment in a 4 × 4 design with 16 plots and built 16 fences, 14 × 14 m, each enclosing 200 m^2 of habitat. The fences either excluded all lizards, excluded only varanid lizards, excluded only small lizards (skinks of several species), or excluded no lizards (fencing controls). They were big and not very pretty. They were also hidden from view and away from public access.

The plots were spread over an area of sandy soil that is long and narrow and covered in a mixture of *Melaleuca* and vine forest. The area is behind a dune ridge at the southern end of a beach that is nearly 8 km long in an area called Ramsey Bay. The area is sheltered from sea breezes and backs onto an enormous mangrove swamp, which then backs onto the Hinchinbrook Channel.

In some ways it was a paradise. After the fences were installed, the first comprehensive management plan for Hinchinbrook Island National Park and World Heritage Area was drawn up, and it designated our site as a Scientific Special Purposes Reserve. The original doctoral student abandoned his project, and I and a local coinvestigator took over more of the management and financing of the project. We recruited some new students, added some unfenced control plots, and started keeping track of more of the flora and fauna so we could examine effects further down the food chain on spiders and leaf and soil invertebrates. It looked like we were set for a long-term project, and we happily settled in.

A typical field trip involved driving from our home base at James Cook University (JCU) in Townsville to Cardwell, about 170 km along the two-lane coast highway. Then we rode in the commercial ferry across the Hinchinbrook Channel and Missionary Bay, down a narrow, winding, mangrove-lined channel, to a small landing where day-tripping tourists could walk on a trail across the dunes and spend an hour admiring the 8-km Ramsey Bay beach and hikers could start or finish their walks on the Thorsborne Trail, a famous 32-km-long hiking track. The owner of the ferry company was supportive of our research and carried us and our gear at reduced rates and even carried fresh groceries from a shop on the mainland. So we lived in luxury, with lots of fresh water, a gas fridge and lights, and a civilized campground.

For 22 hours of most days, we felt that we had the island to ourselves. After an hour or so on the beach the day-trippers were bundled back onto the ferry, and the hikers quickly walked off down the beach to the trailhead. Our tents were set up just behind the dunes. We had a shower tent and could shower sparingly in fresh water. In the morning, we could wander across the dunes and have 8 km of beautiful tropical beach to ourselves, since the ferry with its tourists and backpackers didn't arrive until about 11 a.m. If we were feeling brave, we could go for a swim in the ocean. Although Missionary Bay, 100 m or so behind the dunes, has a massive population of Saltwater Crocodiles, they are *almost* never seen on the ocean side of the island.

Work on this tropical island was not without problems, however. One was how to dispose of human waste. Backpackers on Hinchinbrook at the time were provided with composting toilets at a few designated camping sites, but Ramsey Bay was not one of them. They were allowed to use the cathole technique (dig a hole at least 15 cm deep, do your business, fill it in) when not at those sites. The Queensland Parks and Wildlife Service (Parks and Wildlife) suggested we should either do that or use a Porta-Potti.

Long-term use of a Porta-Potti by up to five people on three- to four-week field trips is a grim business. Even our friendly ferry operator wasn't thrilled with the idea of us emptying it into the toilet on the ferry. We tried catholes, but they quickly reached an impossible density. We were getting desperate when a creative Parks and Wildlife employee pointed out that just over the dunes were 8 km of ocean, uninhabited by anyone, and 22 hours a day, unobserved by anyone. In Queensland, the parks service has jurisdiction above the high tide line, but the marine authorities have jurisdiction below the high tide line.

The solution the Parks and Wildlife employee suggested, which carried us through the next several years with only a couple of glitches, was a simple one: Acquire a folding frame with a toilet seat, a small shovel, and plastic garbage bags. Carry the frame and a bag over the dunes, below the high tide line (and thus out of Parks and Wildlife jurisdiction), look both ways to make sure no errant backpackers or tourists are about, sit down, do your business, put paper in garbage bag to be disposed of along with other paper waste, pick up excrement with shovel, break it up, and throw it well out into the water. This may sound horrifying, but truly, it isn't. Dumping sewage in the ocean works fine when population density is four people per 8 km of coastline. A single dugong produces far more waste than we ever did, and it is a very large ocean; if we had been on a boat with fewer than seven people on board, it would actually have been legal, and still would be, according to Queensland marine regulations.

It did lead to a (very) few "situations," though, when people forgot to look to see whether the toilet seat and shovel were missing before going out to enjoy the morning light or forgot to look for backpackers. One memorable time, someone was just settling into business when a low sightseeing flight passed overhead, then circled back for another look, with everyone waving and photographing out the windows.

Another ongoing problem was insects. With sand dunes blocking much of the ocean breeze, and a vast mangrove swamp 100 m or less behind us, mosquitoes were common, and anyone who was sensitive to sandfly bites quickly became covered in reddish spots about the size of a 10-cent piece. In summer, there were swarms of biting flies of various species, some biting very painfully, some (even worse) painlessly, so the first indication of an attack was the blood running down, often from your face or your ear, as we tended to wear long sleeves and long pants tucked into long socks, even in midsummer in 30+ °C heat.

The only good thing about the insect problem was that, after being swatted, the larger flies worked well as bait. We tied them to a bit of fishing line on the end of a stick, a most entertaining method to catch skinks. One of the small skink species, the Black-Throated Rainbow Skink (*Carlia rostralis*), seldom emerged from under vegetation. They are extremely aggressive in catching flying insects, however. We could swoop a dead fly above a bit of cover, and with luck a skink would leap out, bite down, and hang on for dear life as we swung it in and captured it.

We also dealt with problems with our operating permits. In Queensland, the Parks and Wildlife Service often seems to feel that scientists are annoyances that want to interfere with the proper management of parks. As a result, they can be very difficult about permitting. This isn't helped by the fact that the service likes to consult with local managers, who are very mobile; with each promotion, they move to a new place, so they often have little idea of the history of the places they are managing.

Kat Markey, a volunteer field assistant, using an infrared thermometer to take the body temperature of a Black-Throated Rainbow Skink (*Carlia rostralis*) just after capturing it in an enclosure by "fishing" as described in the text. Photo by Ross Alford.

Initially, I had assumed that, since we were working in a Scientific Special Purposes Reserve that had been designated specifically because our project was there, we should have relatively little trouble with permits. Being written into the management plan should take care of the history problem. It didn't work out that way. Permits last anywhere from 18 months to 3 years, maximum, and almost every time they needed to be renewed, we had a fight on our hands.

Typically, a new local manager would go out to the island for a look around. Sometimes, they would actually decide to check out the Scientific Special Purposes Reserve. When that happened, there was trouble. They were horrified. Ugly, intrusive structures in what was supposed to be pristine wilderness! What if a member of the public encountered them! Unthinkable!

We had put signs in the places most likely to be encountered by anyone who strayed off the marked path from the landing to the beach, explaining the nature of the research, and had never received a complaint. The project was in management plan and was gathering valuable long-term data on the ecology of the interdunal environment. That typically didn't deter the parks people, who almost always wanted to revoke our permit.

One local manager was absolutely adamant. He was shocked and horrified. Our project was a blight on the landscape. That it was part of the management plan was irrelevant, that we had the opportunity to build a unique data set was irrelevant too, since ecological knowledge is irrelevant to maintaining walking tracks and composting toilets. We would have to be out in six months. After some coaxing, he arranged to bring the regional permit manager and meet with us on the island to look over the project and discuss it further.

We drove up to Cardwell and met him, thinking we were all going over with the Parks and Wildlife people on the local manager's official Parks and Wildlife boat. We met them as he was launching it, and it turned out that as civilians who hadn't had the proper inductions, we weren't allowed to ride along. We rushed back and caught the morning ferry. As the ferry was pulling out of the harbor, we were just in time to see the Parks and Wildlife boat stopped in the middle of the harbor, with the local manager, stripped to his shorts, jumping off the back. He had forgotten to put the drain plugs in before he launched, and his boat was sinking.

Fortunately, he got the drain plugs in, the bilge pump worked, and apparently there were no hungry crocodiles in the harbor. The managers joined us on the island, with the local manager in a rather subdued mood. We inspected the pens, showed them our signage and camping area, explained that it was all completely out of public view and that the only comments we had ever had, which were very few, had been curious and positive when people discovered what we were doing. We indicated that of course we would leave the site in its original state when we eventually finished.

The local manager was still adamant; he wanted us to wrap up and take things down and get off the island and was only willing to look at six months. We were asking for three years. We explained that it would take time to understand all the effects, and this was a great opportunity to gain an understanding of the ecology of interdunal systems that should apply to hundreds of islands managed by the Parks and Wildlife Service along the Queensland coast. The regional permits manager, who had the final word, said he had reached a decision; he would give us 18 months, and after 12 months we would reinspect the system and look at our permit returns and data analysis and renegotiate.

I felt a wave of relief that our large investment of time, money, and student futures wasn't going to collapse in a heap. I thanked the regional manager and assured him that we would take full advantage of the time and demonstrate that the project was worth continuing. The local manager was clearly unhappy and said so. My JCU coinvestigator turned to the local manager and, in a moment that horrified me, said, "Fortunately, that won't matter—in a year, you'll have been promoted and be somewhere else anyway."

I had a brief vision of the local manager hanging on just to deny us our permit, but fortunately, it didn't happen. The project continued, students finished, more students started, and the project might still be going to this day . . . except for other problems that finally killed it.

First, Parks and Wildlife decided to build an elevated walkway past half of our plots. They were worried about erosion caused by tourists walking to and from the ferry landing and that the beach might eventually cut the dunes to the point at which an extreme weather event would erode through them and cut the northern part of the island off from the southern part. The walkway would lead tourists to an area where they could cross at a natural rocky region less likely to do that.

The walkway would mean that our plots would be exposed to public view; they were not pretty and would have to go. In the meantime, nature intervened. A category 5 tropical cyclone, Yasi, passed through the study area on the night of 2–3 February 2011. It irreparably damaged the fences and severely disturbed the vegetation. A few months later, we paid to remove the debris, and our experiment in paradise finally ended.

In the end it was all worthwhile. Two people finished doctorates on the island; two others finished bachelors of science (with honors) degrees—a research degree with a publishable thesis. Two others did projects leading to parts of masters' degrees. Many volunteers got a taste of the joys and pains of fieldwork. Hundreds of glorious tropical sunrises were seen over 8 km of nearly pristine sand, and hundreds of late afternoon breaks were taken in the sea breezes on top of the dunes, relaxing after hard days of data collection. Several papers have been published thus far and several more are still in progress.

About the Author

Ross A. Alford grew up in South Florida, where he divided his time between roaming outdoors and reading natural history and science fiction. After undergraduate and MS degrees in Gainesville, FL, he did a PhD at Duke and got a job at James Cook University in Townsville, in tropical Australia. He has never looked back. He has lived in the tropics (19° S) for 37 years, traveled over most of Australia, and spent at least a year and a half living in tents. He has made major contributions to the study of invasive species, animal movement, amphibian ecology, and amphibian declines. His achievements were recognized by his promotion to the position of professor and personal chair at James Cook University less than 20 years after his initial appointment.

38

ISLAND CASTAWAYS AND THE LIMITS OF OPTIMISM

Alessandro Catenazzi

The sun woke us up. We were wet. Fine white pearls rolled along our eyebrows, hairs, and skin. They formed small puddles around the eyes and the navel. The morning dew carried in the salty tears of the ocean, and as the sun rays broke through the fog, new hopes of returning to the mainland that day.

At dusk the night before, we had left the island's shore where sea lions howl all night, stalked by insatiable vampires, and hiked up the narrow canyons dotted by nests of Inca Terns and swifts, to reach the peaks where grasses and tillandsias plucked the tiniest of water droplets from the passing fog. Thirsty, after yet another day spent waiting for good-natured fishermen, or anyone, to deliver us from this arid land, we ascended to this temporary home for a second night. There was no need to drink, as the air at the top of the island was so saturated, inebriating at the end of a fruitless day of waiting. We laid down our wool blankets on the red sand and satisfied our hunger with tenderness. In this best of all possible worlds, our island was the finest of islands, and we the best of all possible baronets.

The research at the Paracas National Reserve, Peru, had taken a sharply descriptive turn after it became evident that an experiment manipulating the amount of beach wrack (algae and other marine organisms cast up on shore) would have required far greater resources than those available. I originally planned to conduct the experiments on the mainland, but after witnessing the experiment's excruciating failure, the appeal of nearby islands, where biological phenomena of interest were amplified, became too great to resist. The idea of my dissertation was to compare lizards' use of marine resources across segments of mainland beach where I controlled the amount of marine wrack accumulation. Seaweed cast on shore attracts flies, beetles, beach hoppers and other crustaceans, all potential prey of lizards. Initially the plan had been to remove the wrack manually, mostly green algae and kelp, more arriving every day courtesy of high tides. It quickly became evident that approach was futile. Mounds of algal accumulation outpaced my ability to remove it. I thus decided to build algal nets, which like barricades against the Pacific would trap floating algae and prevent their accumulation on land.

As much as I put tremendous hope into my sea walls, their capacity to trap algae was limited during calm ocean days and disastrous when high tides and choppy waves delivered their loads. There were enormous amounts of algae on certain days, attracting hordes of Peruvian Lava Lizards (*Microlophus peruvianus*), including on experimental plots that were supposed to receive no wrack. The lizards were ecstatically running down to the tide line, dancing in ballets choreographed by the incoming and receding waves, grand jetés to catch kelp flies that mocked and ridiculed my vain efforts. To add literal oil to the fire, a rock pierced the oil pan of my car as I frantically drove bags full of pestilent and decomposing algae away from my wrack-free (if only!) plots through the desert. A tow truck had to come to the reserve's parking lot to rescue my car and tow it to the city. During the ride to the mechanic workshop, as I was sitting in the truck's front seat and making small talk with the driver, I had the time to explore all my misgivings about time lost and decided to abandon the experiment and, instead, pursue a comparative observational approach.

I had, by then, spent several months in the desert and noticed that Leaf-Toed Geckos (*Phyllodactylus angustidigitus*) were far more abundant and easier to catch than lava lizards. Geckos lived both along the coast and in plantless hills far away from the ocean, in contrast to lava lizards, which aggregated patchily along beaches. The gecko's ubiquity opened opportunities to examine how their diet, demography, and distribution varied according to the type of coastal landscape and availability of marine resources. Like geckos, other nocturnal animals such as solifuges and scorpions occurred throughout the desert. Scorpions have added benefits as study animals because one can easily spot them by flashing the ground with a black light lamp and because they slowly digest their prey externally, facilitating dietary studies. I tested the regurgitation technique on geckos, and I quickly succeeded in developing a protocol to recover stomach contents without harming the geckos. Small clips of the tails, which the geckos regrew, guaranteed that I could complement dietary analyses with stable isotopic studies, which can trace marine contributions to terrestrial ecosystems.

The next step was to select and mark study plots in the coastal desert. The criteria were based on availability of marine resources, either because of the type

The Narrow Leaf-Toed Gecko (*Phyllodactylus angustidigitus*) feeding for beach hoppers and flies among sponges cast on the beach of Paracas Bay of southern Peru. Photo by Alessandro Catenazzi.

of beach—pebbly and shelly beaches retain the most wrack—or the type of food, varying from beach-cast wrack to carcasses of seabirds and sea lions and their scavenging arthropods. A rugged coastline with tall cliffs is an obstacle to Leaf-Toed Geckos, which unlike other geckos cannot stick to vertical surfaces. They must walk down to the beach and feed on intertidal organisms. Once I selected plots, with three replicates for each type of coastline, I started marking the corner of each plot with steel rods. Because of the strong winds, called *paracas* (literally "sand rain" in Quechua), I thought I needed something sturdy to delimit the boundaries of my plots. As desolate as the desert may look on some days, peering eyes were scouting out my bizarre rebar planting behavior for free construction material, and it didn't take long before most rebars disappeared from plots. Another setback.

At the time, I was living at the reserve's headquarters, helping with cooking, upkeep, and maintenance. One of the brooms used for the long daily routine of removing sand from the interior spaces had just broken, leaving me with the stick in my hand. I had the images of the nearby Nazca lines in my mind, giant figures and lines drawn in the desert, which require a bird's view to appreciate fully. The two thoughts combined into the idea of drawing small trenches in the desert as a way of marking my plots. The following days, I started walking along the full perimeter of each plot, dragging the decapitated broomstick into the desert soil. Not quite like *The Adventures of Priscilla*, costume trailing from a perch atop the roof of a bus blazing across the Australian Outback, but as close as you can get to that image. After marking the external perimeter, I subdivided and marked each 250 × 250 m plot into smaller 50 × 50 m plots to facilitate geolocation of the geckos and scorpions that I would capture and recapture over the course of my research. I waited for the next paracas, and it was with great relief that I witnessed the resilience of this simple and disturbance-proof marking method.

The paracas were not the only challenge to the marking system of my plots. Curious about the sudden apparition of the squared-shaped patterns, some local people conjectured about planned real estate and luxury hotel projects, landing pads for extraterrestrials, or archeological study areas. Getting to one of the plots required ascending a red hill, colored by a type of porphyritic rock, near places where archeologists and *huaqueros* (graverobbers) had previously excavated and found ancient mummies and their precious textiles. Walking to the top of the hill at night, under clear skies and the moon's reflections on the deep red igneous rocks, was a nearly mystical experience. Disruption came when unknown and unscrupulous huaqueros suspected that the plot markings indicated the location of burial sites. They visited the plot on a dark new-moon night, when we were busy working in another area of the desert, and started digging at various locations within the plot. We discovered many holes in the ground the next time we visited the plot. It must have been a long night of disappointments to discourage the diggers and finally convince them to abandon the site.

Initially, all study sites were on the mainland. The reserve was employing volunteer rangers, and one of them decided to remain at the end of the program and help me with my research. We built up a routine for our nocturnal meandering through the desert plots. During the drive, we would listen to a radio program focused on helping people with mental health issues—they would call the station, and the host would talk through their concerns with her soothing voice. Sometimes we had chips and drinks at sunset, looking at the sun sinking into the desert and ocean behind, saturating the hues of yellow, brown, gray rock layers and sand. The research data were slowly telling a story of extreme reliance on marine nutrients: some geckos, for example, fed nearly exclusively on beach hoppers. Their main daily movements were nocturnal walks from desert burrows to the beach and back.

Like the geckos, our diet was becoming increasingly maritime. The rangers periodically inspected unauthorized people caught fishing illegally, or collecting scallops, other mollusks, and seafood from protected areas. The seized seafood was then given to clubs of mothers (many of whom were the fishers' wives) in nearby towns, but when transportation was not available, we had to consume the seafood before it spoiled. Whenever some fishers returned to dock too late in the morning, found no taxis waiting, and feared their catch spoiling, I would rescue their day by driving them to the nearest fish market, in exchange for getting several kilos of fresh fish. I would tell them about my research, and our nocturnal wanderings, but I suppose I did not meet a sufficient number of the fleet, as I discovered one night.

The night we unintentionally terrorized fishers started like any other night. We were working at a desert plot near the dock. We had just completed the surveys of geckos and started our scorpion survey with a portable, meter-long neon black light. We were wearing light beige, almost white, pants and carrying the lamp at waist height. The night was very dark, with the moon just a slight crescent, and the beach was populated by dozens of scorpions. We were fully absorbed by our work, unaware of how the reflection of the blacklight lamp would appear from far away. Two animated pairs of torso-and-head-less legs appeared vividly to any passerby in the depth of the night. When taxis carrying fishers started arriving around 2 a.m., we saw the cars approach at high speed, reluctantly slow down as they neared our plot, then stop entirely, and finally frantically make a U-turn and speed away. It must have looked like the day of reckoning had arrived.

We made good friends with several of the fishers, especially once we started visiting nearby guano islands, and it might have been karma that left us stranded and thirsty on the island of the sea lion rookery, payback for the terror we briefly inflicted on the fisherman guild. The islands had even more astonishing examples of terrestrial animals relying on marine nutrients. The geckos and scorpions were feeding on seabird ticks and scavenging beetles found on sea lion carcasses; lava lizards went after the red feces of sea lions, which had consumed shrimp; and vultures flew over dead pups and placentas. After two days spent without water, we eventually decided to call for help. I had one of those brick-sized cell phones, with a retractable antenna, an early low-cost model marketed by *Telefónica* during those years. It was esthetically and practically very unpleasant to carry but had the benefit of being able to get the feeblest of receptions. We called the reserve, and their boat promptly came to rescue us. As we were surfing the waves back to the mainland, my mind returned to the two nights spent on top of the hill, and I daydreamed about waking up every morning, laying in the midst of the wildflowers of *Tillandsia* and *Solanum*, and being a gardener of that little plateau, so close to the sky.

About the Author

Alessandro Catenazzi is associate professor at Florida International University, where he received his PhD in 2006 and now leads a conservation biology laboratory. He works on the systematics, ecology, and conservation of amphibians and reptiles. His research documented the collapse of Andean frog communities following outbreaks of chytridiomycosis, explored the effects of anthropogenic factors such as prescribed fire or riverine disturbance on endangered frogs and lizards, discovered more than 60 new species of frogs and lizards, and contributed to the creation of protected areas in the Andes–Amazon region. He lives in Miami with his two Central American expat sister cats.

39

LESSONS IN PATIENCE
Frog Eggs, Snakes, and Rain

Karen M. Warkentin

In my memory of 1992, there was no rain at Sirena Biological Station, in Corcovado Park, Costa Rica, until the 19th of July. My journal reveals that is wrong. I noted sprinkles, drizzles, and even 22 mm one day, none of which filled the frog ponds. In that rainforest—where 10 cm of rain often fell in a day and about 5 m in a year—it took a serious downpour to get the frog reproductive season started. My memories of rain are emotionally aligned with the frogs' needs.

I had been to Sirena a year earlier, on a field course, looking for an interesting frog and a research question that would let me live in the rainforest, studying eggs or tadpoles for my doctorate. At night, I had seen Cat-Eyed Snakes (*Leptodeira ornata*) eating the beautiful, transparent eggs of Red-Eyed Treefrogs (*Agalychnis callidryas*) hanging on leaves over a pond. I had also seen a few embryos hatch, after I accidentally bumped an egg mass. It was quick enough that I began to wonder if they might be able to escape from egg-eating snakes. Back in Texas for classes, I spent months discussing my "escape-hatching" hypothesis with other scientists, some intrigued and others dismissive. As the field season approached, I carefully planned my experiments, hoping to see snakes try to eat embryos that might be able to hatch. Did I have a novel idea worth testing and the start of a dissertation project? Or just an overactive imagination and no tractable research plan? The snakes and eggs would tell me.

Corcovado journal excerpts are edited for conciseness, clarity, and relevance.

I knew the *Agalychnis* pond in Sirena had enough egg-eating snakes for my work, but watching them feed would be another matter. The sometimes-rapid movements of snakes belie their immense capacity for stillness. Even Cat-Eyed Snakes, actively searching the vegetation for eggs, spend long periods immobile. On the field course, standing hip deep in the pond in the rain at night, waiting for a snake to move, I had wondered if I really had enough patience to work with them. I knew I would need new snake-watching skills to test my hypothesis, but when I returned to Sirena I encountered an unexpected challenge. The "pond" was dry. No place for frogs to breed meant no eggs for snakes to eat and no way for me to answer my questions. Still, it was 18 June, so it should rain soon. Expecting a downpour any day, I set up my handcrafted snake cages and prepared to start my research as soon as there were eggs.

Sirena station housed park staff, biology students, and ecotourists amid the amazing wildlife of the largest protected patch of lowland Pacific rainforest in Central America. My tent was pitched in the unwalled second floor of the main building, under a tin roof that would amplify every raindrop—if rain fell. I woke each morning to a raucous chorus of Scarlet Macaws drawn near by a young bird, rescued from poachers, who still roosted in the building. At first, I delighted in long walks, getting to know the forest, beach, and intertidal. I saw so many crabs— robust land crabs that emerged from burrows to wander the forest floor at night; hermit crabs in their mollusk shells, clustered on fallen fruit at the top of the beach; alert orange fiddler crabs scuttling to their burrows when I approached. As the rainless days continued, I thought about studying crabs instead of frog embryos. I could have been working, testing a hypothesis already, like the other biologists. To feel like a scientist, I helped count and measure palms, photograph lizards, watch butterflies, making tiny contributions to other people's work.

Still, I was not alone in my frustrated desires. The hungry snakes and sexless frogs shared my powerlessness to affect the one thing that mattered. When would it rain? My Corcovado journal reveals our shared stress and my peculiarly human responses.

Sirena Station, 6 July 1992—Rain structures our lives, here in this rainforest. Or, at the moment, lack of rain. Lack of rain keeps my frogs up in the trees, males calling futilely from dry ponds, snakes first searching and now only waiting for eggs that don't come. Lack of rain maintains this oppressive heat and my constant sweating, even seated immobile, only writing. Here, we become more superstitious. So much of our lives are beyond our control, at the mercy of the forest and its creatures. We develop rituals that allow us to believe that we have some control, that what we do matters. We joke about the mosquito gods and the cult of the coil. We joke too about the rain, but still we leave our laundry out and go without umbrellas, hoping for rain, or always carry umbrellas, hoping for lack of rain. When by chance the rain fits

our superstitions we point it out. "You left your laundry out? Of course it's raining."
In reality, the rain gods pay us no heed, preoccupied with their own agenda in this El
Niño year. I repeat my laundry-hanging rituals in vain, hoping for the half-bucket
rain that will bring my frogs down from the treetops, waiting for a chance to work
in this place of tourist holidays.

As Sirena continued to be—by egg standards—rainless, I lived with uncertainty. Could the embryos really be capable of assessing risk and fleeing from
danger? It seemed far-fetched but also logical, among so many hungry snakes.
Or, was I just wasting time and resources, as some had suggested? No eggs meant
no data and no answers. To fill the waiting, I left for a few days, hiking inland
across the park.

Los Patos Station—The red mud has only enough moisture to be sticky. Thick and
hard it catches my boots, snaps sounding at every step as I break its grip on my feet.
We walk all day in the heat, sweat sliding down our surfaces, soaking our clothes and
starting to ferment. Sometimes I wipe my arms with a bandana. It gives a moment's
coolness, a few seconds respite before the beads congeal into one smooth layer and
I stay wet as in a bathtub, my skin softening and wrinkling in its own water.

At 1:30 it rains. The first drops fall on the canopy and are consumed by foliage,
reaching us only in sound. Fifteen minutes later, water arrives to the ground and its
creatures, dripping through layers of leaves, accumulating on drip-tips, splattering
on the trail. My sweat is diluted by the splatters and, somewhat less than body tem-
perature, they cool me. I relax in the knowledge that everything in my pack that is
still dry is sealed in Ziplock bags. (I don't think about the clothes molding in two days
of sweat and mud.) Rain soaks the leaf litter but the red mud resists its passage. Only
the top layer softens to slickness, water accumulating in red puddles on the trail. On
the hills it flows from footstep to footstep in red waterfalls. Puddle, waterfall, puddle,
waterfall, I make my way. The splashes raised by the hiker in front of me will be
permanent red stains on my off-white field pants.

When I returned to Sirena, the ponds were still dry. Advised that "a watched
pond never fills," I left again, walking northwest along the beach to the amazing
Llorona Plateau, home to the largest trees in Central America.

Llorona Beach—The beach is covered in orange crabs feeding on intertidal sand,
two pointed eyes on stalks watching warily as I approach. The occasional wave that
comes farther up sends them scurrying, concentrating a line of more intense orange
along its front. A splendid sight. I think of fiddler crabs as having rounded eyes, the
stalks bulbous on the ends. These creatures have pointed eyes, topped with a bright
orange spike, a sort of clown eyebrow for crabs. It gives them a perpetually surprised
expression. Walking here last night, I was expecting crabs, the whole beach littered
with lumps of crab-chewed sand. The walk was beautiful: wide wide tidal flats, soft
wet sand glistening with the moon's reflection, the occasional plover or sandpiper

flushed unseen, piping before us, and the white edges of the waves breaking far out to sea. Higher on the beach, before the Rio Corcovado, we crossed cat tracks. First two sets together, large and round enough to be jaguar. These were fresh, and so close to the falling tide line we must have been within half an hour of their makers. Probably our paths crossed in the dark, and they saw us but not we them. Later we saw even bigger tracks, a solitary individual, and a set of smaller ones, probably margay. It seems the beach is full of cats.

Returning from the wonder of Llorona to dry Sirena, I was tired, blistered, and miserable. Was my egg project doomed? Should I try to study crabs? The frogs and snakes had no choice and, with them, I continued waiting for rain. Then on 19 July our luck changed.

Sirena, 19 July, ~10 p.m.—Perhaps, finally, the days of half-bucket rains are upon us. It started raining, really, at about 5:30, and it's raining still. I emptied the rain gauge at 9:45 and it was already 88.2 mm. I've been saying 10 cm is what we need to fill the ponds and get some frog action happening. And it looks like we've got it. The drainage ditch across the airstrip has water, as do the grass flats. The streams are up, the Naranjo-beach pond is hip deep, and my pond is gathering water and frogs rapidly. The frogs were just bouncing in as I sat there. Finally! It's been a long dry season.

The wet places of Sirena were a frenzy of frogs, an orchestra of nine species calling, an intense convergence of mate-seeking females, laden with eggs, hastily migrating from forest to chorus.

Wading in the ponds, I am reminded of the joys of being wet to the crotch in rubber boots full of water and tapir shit, of shivering, insect bitten, coughing from breathing in bugs attracted to my headlamp, while trying to keep track of amplectant pairs. Yes, this is what I am here for! This is the true happiness of a crazed herpetologist.

*12:30 a.m.—The rain had quieted down but now is back with a vengeance. We must have 15 cm. So much water is blowing around that my tent, under the station roof, is getting wet. There was a drip, a splash onto my face. Perhaps this awakened me. This or the thunder. Finally it really **is** raining. Finally I have opportunity and reason to go in over the tops of my boots. This thunder is music to my ears. The splatters of rain into my tent are kisses. I feel great vicarious delight in the frogs' good fortune. It has been a long dry season.*

And it rains, it rains. The lightning and thunder are almost continuous; the building shakes and I smile. This is it, the turning point. This morning I woke, after two days of gloom and despair, in an excellent mood. And for no reason, I thought. I just felt, no matter if the rain is taking its own sweet time, it is right for me to be here; I am learning things. And now, the rain comes. Is it because I loosened my desire? Was my inexplicable happiness simply a foreshadowing of this wondrous night, finally wet again? It does not matter. What matters is only that it rains, frogs call and mate, and the life cycles of the wet ones continue. I am privileged to be here, to see it. Life is good. The frogs and I are enjoying this night. They out in the wet and me, no longer shivering, in my tent watching the play of lightening in the sky and seeing again the plop, hop, splash of frogs coming out of the forest to the pond, frog by frog by frog. The rain is so loud it drowns out the sound of their calls. Amazing.

This night at the Agalychnis *pond, my pond, the snoring chorus of Milky Tree-frogs was so loud I could only hear the* Agalychnis *in the lulls between snoring. I don't know if the* Agalychnis *only call in the lulls or I just cannot hear them over the snores. Their little clucks are pretty quiet. But they were there first. They have been waiting a long time for this night.*

The next night *Agalychnis* festooned the plants with egg masses, and I began my life of repetitive egg counting. At the pond, I checked clutches twice daily until all the embryos disappeared, hanging cups underneath to catch hatchlings and determine how many eggs survived to become tadpoles. In experiments, I used cages with removable dividers to control when snakes could reach eggs and observe predation on late-stage embryos. The snakes continued to test—and train—my patience. I learned to move slowly and silently, working in dim light, to check for snake activity and count experimental eggs. I learned to count eggs often until a snake awoke—some soon after dark, others not until 3 a.m.—for a baseline, then take extreme care, freezing if the snake moved, until it began feeding. Once they had a mouthful of eggs, the snakes seemed to shift focus and would continue feeding while I watched. After I spent many nights sitting in near darkness, watching them eat, some would continue feeding through cam-era flashes. This gave me the first photographic evidence to support what my egg-count data also showed: at later developmental stages, Red-Eyed Treefrog embryos *do* hatch in response to snake attacks, and most escape. This became the basis of my doctoral dissertation and my entire research career.

Twenty-four years later, in Panama with students, I set up a wasp-feeding station for an undergraduate to video-record escape hatching. Walking slowly through the nearby pond, holding a leaf with young eggs and looking for egg-eating wasps to recruit for our experiment, I encountered a lovely green Parrot Snake (*Leptophis ahaetulla*). This normally skittish species is the only snake we know that eats Red-Eyed Treefrog eggs in daylight. This individual seemed unusually calm, so I started recording video; it did not flee. I extended the eggs-on-a-leaf, then held as still as I could. The snake tongue-flicked and slowly approached as my student watched. The video shows a close-up of the snake eating eggs from the leaf in my hand. After a few minutes, under the sound

Karen Warkentin checking and counting eggs in a Red-Eyed Treefrog (*Agalychnis callidryas*) clutch at the *Agalychnis* pond, Sirena Station, Corcovado National Park. Photo courtesy of Erika Deinert.

of birds and insects calling, there are quiet human voices. First, scientifically, discussing the snake's feeding behavior. Then, my understated wonder: "This is pretty amazing. I've never hand-fed one before." "Really?" "Yeah." "Wow, I thought this was like your millionth time." "No." This was unique, but even after 30 years working with egg-eating snakes, every snake encounter in the wild feels like a gift. The rain and the beautiful eggs it brings are gifts, too, that the snakes and I share.

About the Author

Karen M. Warkentin is a Canadian herpetologist, a professor of biology and of women's, gender and sexuality studies at Boston University, and a research associate at the Smithsonian Tropical Research Institute in Panama, where they have studied environmentally cued hatching since 1998. From 1991 to 1995, as a PhD student at the University of Texas–Austin, they spent 24 amazing months in Corcovado National Park, Costa Rica. Their research has been funded by the Natural Sciences and Engineering Research Council of Canada, the National Science Foundation, USA, the National Geographic Society, and the Smithsonian. For more information, see their lab website at https://sites.bu.edu/warkentinlab/.

40

SOUNDS OF SILENCE ON THE CONTINENTAL DIVIDE

Emily Taylor

Frogs and other amphibians are an enigma to me. My training has given me expertise in what makes the scaled creatures we know as reptiles tick, but frogs? I learned about them a bit as a herpetology student, and then I figured out a little more from teaching my own herpetology classes. However, I live in California where snakes and lizards dominate the landscape.

I heard stories from colleagues about wet and wild places in the world where frogs thrived, so much so that you could holler at the top of your lungs right next to someone and they still couldn't hear you. My graduate program offered a semester-long tropical ecology course, but my research advisor told me I had too much work to do in the desert. The tropics would have to wait.

In 2009, as a young assistant professor, I was invited to the Continental Divide in the Panamanian highlands to help a colleague plan their field biology class. Finally, I would have the opportunity to experience the tropics in all their glory: from the silent snakes to the bellowing frogs; from the sweet mountain air to the savory food; from the clear streams to the landscapes cloaked in green. The Continental Divide is a mountainous area that runs through the middle of Central America and gives rise to and separates the watersheds on the Pacific and Atlantic sides. Lush and wet, this area famously houses an incredible diversity of wildlife, from insects to birds to frogs.

What I didn't know was that I was going to bear witness to the fallout from a crumbling tropical community due to a deadly fungus that had arrived five years ahead of me.

We traveled to an extremely remote national park officially named Parque National Omar Torrijos after Panama's beloved leftist dictator of the 1970s, but often called El Copé after the nearby town of about 1200 people. The dirt road from El Copé to the park was wet, slippery, and pockmarked heavily with the tracks of other four-wheel-drive vehicles that had made the journey. When we emerged from the vehicle, I found myself in a small parking lot next to thick primary forest, completely shrouded in clouds. It wasn't exactly raining, yet moisture dripped from the air in that peculiar way of cloud forests. We spread out through the forest to look for creatures.

Verdant vine-wrapped trees stippled with epiphytes enclosed us in a solid blanket of that healthy rainforest smell that I wish could be bottled up and used as a home fragrance. The fungus-rich soils beneath my feet sprang back after I stepped, leaving no trace of my tracks. Moisture dripped from the foliage to join the streams that looped around the trees, pooling here and there before making its gradual way down the northern slope where it would join the Belén and Coclé del Norte Rivers before emptying into the Caribbean Sea. Omar Torrijos National Park was the picture of a perfect, healthy cloud forest.

Except for one major thing.

The forest was completely silent.

Prior to 2004, the pools and streams would have been crawling (or swimming, hopping, jumping, or other froggy word of your choice) with amphibians. Tiny long-legged bodies would have scattered under our feet as we stepped through the leaf litter. There would have been an ear-splitting cacophony of male frogs singing their hearts out to attract females.

But most of those frogs had died five years earlier, resulting in the heartbreaking collapse of an entire amphibian community due to a microscopic fungus that had swept north to south through Central America in a wave of destruction.

No more Giant Glass Frogs (*Sachatamia ilex*) showing off their internal organs through transparent skin.

No more Clown Frogs (*Atelopus varius*), whose territorial battles start when males wave at one another because the gurgling water drowns out their squeaklike vocalizations, and whose females stomp on males who invade their territories.

No more of at least twelve different species of Brittle-Belly Frogs (*Craugastor* spp.), which skip the tadpole stage and develop directly into little toadlets.

Tropical rainforests are complex ecosystems, where species are connected to one another via fragile threads that are invisible when the system is healthy. However, pull one thread, and the entire web can fall apart. In the five years since the fungus had hit, how had the community of Omar Torrijos National Park been affected by the mass die-off of frogs?

I got my first answer when someone hollered "Frog!" I crashed through the forest toward the voice, ducked under a thick vine, and splashed into a stream . . . but I didn't make it across. As soon as I stepped into the cold water, my boots slipped right out from under me and I was on my back.

I had expected finally to learn all about frogs during my trip to Panama. And learn I did, but not in the way that I had expected. Falling unceremoniously into the stream was my firsthand experience of one way in which amphibians play an important role in the environment. Ordinarily, hordes of tadpoles nosh away on algae with keratinous little beaks. After the fungus arrived, the collective loss of so many of these nibblers led to the gradual buildup of huge, slippery slabs of algae on all underwater surfaces, just waiting to gleefully flip unsuspecting biologists onto their backs.

Sputtering, I hoisted myself up and out of the water, shook myself off like a dog, and ran over to where a flashlight was trained on a small rock on the ground. Sure enough, a little frog about the size of a silver dollar sat shining in the beam, its big eyes peering up at us, its toes flared out to form enlarged round pads.

"A treefrog?" I inquired.

"Looks like a *Smilisca sila*," my colleague confirmed.

The Panama Cross-Banded Treefrog is an extremely common frog, ranging from southern Costa Rica through most habitats in Panama down to central Colombia. This is one of the few species that survived the fungal assault. I kneeled down and gently cupped the little frog into my hand. As I positioned it this way and that to take photos, it tracked me with its big, black pupils, which were surrounded by brilliant, glittery gold flecks.

As I placed the treefrog back onto its perch for more photos, I heard someone call "Snake!" in the distance. This got me moving again. Searching for snakes in a tropical rainforest is like looking for a needle in a haystack, except you're never sure if the needle is even there to begin with, the needle can move around, and sometimes the needle is highly venomous.

Adrenaline pumping, I raced through the forest, slowing only to carefully plod through streams to avoid another dip, and found a colleague with their flashlight trained on a patch of vegetation about head height.

"Where is it? I don't see it!" I squeaked.

Laughing, my colleague held out one of the vines. But it wasn't a vine, it was a snake!

I was looking at a Cope's Vine Snake (*Oxybelis brevirostris*), a true marvel of camouflage, stealth, and beauty. As you might guess by its name, this snake is a vine mimic, its long, skinny, emerald-hued body allowing it to literally disappear into mats of vegetation. It glared at me with the ultimate snake side-eye as it hoisted the front half of its body into the air, effortlessly weightless in its search of a branch on which to mount its escape.

"Okay, Okay, you can go back to bed." I arranged the snake in the tree to snap some photos, and at the first chance, it glided silently into the thick canopy.

I later found out that vine snakes have always been one of the most common snakes in the park. However, locals mused that the vine snake didn't seem to be as common anymore as it used to be. In fact, in the five years since the sudden loss of frogs, most snakes seemed to have declined in the park as well.

What would frogs have to do with snakes? Well, as was the case with the stealthy vine snake, many snakes in the park eat primarily or even exclusively frogs. Pulling the frog thread led to the unraveling of the entire food web. When the frogs died out, so did the snakes, or so the hypothesis went. Given how hard it is to actually find tropical snakes even when they're abundant, it took 10 more years until the tricky math was worked out to document declines in secretive species like snakes. In 2020, a study was published in the journal *Science* showing a crash in snake species diversity in the park following the 2004 amphibian decline. Vine snakes made it through intact, but those encountered by scientists were typically even skinnier than their natural svelte state.

Why did some frogs and snakes survive, while others went extinct? Why have some species recovered since the first wave of fungus arrived, taking advantage of the decrease in competition, while others exist now only in captive collections where zookeepers fastidiously breed them in the hopes that their distant offspring may one day be released back into the wilds of El Copé and other tropical landscapes devastated by the chytrid fungus?

I was merely a tourist to this remote rainforest, but my colleagues toil every year in search of answers to these questions. As I packed my luggage to return home to California, it occurred to me that the biologists fighting for the survival of these meek forest creatures are true heroes. The amphibian declines that began in earnest in the 1980s were a classic canary in the coal mine scenario, acting as a warning to us all about what could happen to other wildlife. More recently, scientists all over the world have sounded the alarm about the collapse of other wild communities, from insects to coral reefs to birds. In some cases, recovery efforts based on sound science have worked: reintroducing wolves into Yellowstone restored the elk-chewed landscape; beavers reintroduced to English forests improved soil quality and reduced flooding; the Leonardo DiCaprio Foundation and Fundación Rewilding Argentina have invested millions of dollars into restoring a fractured landscape in Argentina by reintroducing tapirs, peccaries, and giant anteaters with plans in motion to add jaguars back to this rich area where they historically thrived. Even at El Copé, the recovery of certain frog species is slowly restoring the ecosystem.

A disease had silenced the forest I visited in Panama, a scenario mirrored in hundreds of other forests and wetlands around the world. But loud are the conservation biologists and zookeepers who tirelessly advocate for the frogs that once made these forests ear-splitting cacophonies of life. Because of their efforts, future visitors to El Copé might have wildlife encounters beyond my lonely tree-frog and vine snake and might experience the forest as it once was—a verdant, pulsing hotspot of chirping, beeping, and wailing sirens made by dozens of species of vocalizing amphibians calling their hearts out, as the snakes slink among them in search of lunch.

About the Author

Emily Taylor got a bachelor's degree in English from University of California–Berkeley and a PhD in biology from Arizona State University. She is a professor of biological sciences at the California Polytechnic State University. She has won several awards in herpetology, including the Meritorious Teaching Award and the Margaret Stewart Award for Excellence in Herpetology. If she is not in the field or lab doing research on the environmental physiology of lizards and snakes with her students, she is likely out rescuing rattlesnakes from people's yards or otherwise plotting how to help reptiles reclaim the planet.

Part VI

THE PEOPLE WE MEET, THE FRIENDSHIPS WE FORGE, THE STUDENTS WE INFLUENCE

41

WHY DO I DO WHAT I DO IN THE FIELD?

Joseph R. Mendelson III

People assume that the wildest field stories will involve a viper, a jaguar, or some horrid mosquito-borne disease. Nope. The most interesting stories revolve around the people one meets while doing our work. Some encounters are wonderful, some are hilarious, and some are terrifying. While romping around remote—and not so remote—areas looking for amphibians and reptiles seems perfectly normal to people like me, to a nonbiologist this is a profoundly weird thing to be doing. We also tend to work late at night, which only exacerbates our weirdness. Many people fear some of our subjects, and most people pay no attention to the innocuous ones. So, a good amount of time is spent explaining oneself to people encountered along the way. Add a language barrier in foreign countries or a deep cultural gulf between the field biologist and local people, and miscommunications can arise.

I suppose I was preadapted to a lifetime of odd encounters with people, given that they started within my own family. There is nothing pleasant about road-killed animals, but they can be valuable as research specimens. So, naturally, I started saving these treasures in the family freezer for the local natural history museum. My mother once thawed out a large, smashed Red Diamond Rattlesnake, having mistaken it for a roast beef. Upon discovery, she abruptly declared, "We're going out for dinner tonight!" The next day she taped off a section of the freezer for my exclusive use. I had very supportive parents, but at dinner I got the question: "Tell us again, why do you do this?" I've now spent a career responding to that same question.

The strangeness of our pursuits is magnified when visiting other cultures. When poking around the outskirts of villages looking for animals, in Mexico for example, it is polite and smart to spend considerable time introducing yourself and trying to explain what you are doing to local folks. Like their rural counterparts everywhere, these folks usually consider amphibians and reptiles with antipathy or utter disregard. The communications challenge is not simply explaining what I am doing but convincing them to believe me. Why would a *gringo* travel all the way to their remote corner of the world, brave hordes of mosquitoes at night in the swamp that nobody enters, just to look for frogs? After years of trying, and generally failing, to provide an answer that made any sense to people, I settled on a storyline that seems to work. I explain that I am a teacher in the United States and I want my students to understand something about the beauty of their country, but I can't bring all the students with me. Instead, I spend my summers collecting samples to share with them so they can compare them with the creatures they know, and they learn something about both countries. This is actually very close to the truth, so I'm happy with my strategy. Before I came up with my storyline, however, I had some mis-steps.

In the 1980s, Guatemala was mired in a horrific civil war, and I ventured right into the middle of it to look for toads. In my youthful ignorance I didn't realize that the US government had played a foundational covert role in starting this decades-old conflict. Most of the locals around the coffee farm I frequented clearly were spooked by my presence. The few that would speak to me simply refused to believe that I was just looking for toads and snakes. I slowly pieced together that everyone assumed that I was a CIA operative with a ridiculously unbelievable cover story. That trip culminated in a terrifying encounter with armed villagers in the middle of the night. My conversations at this study site never got to why I was doing my work because no one believed what I was doing was real. Imagine a US citizen going to Afghanistan today and wandering around catching lizards, and you'll see the folly in my approach.

Another communications failure involved a Mexican colleague and myself making the poor decision to climb over a fence late at night to survey a patch of remnant cloud forest in Veracruz, Mexico. Rule 1 of field work: never cross a fence without permission. Rule 2: never survey an area at night without first scouting it out during daylight and introducing yourself to the locals. Jaguars and vipers are not the biggest threats. Spooked locals who cannot possibly guess why two people with headlights are wandering around their property in the middle of the night are the threat. We succeeded in finding the lizard that my friend needed, so it was a great night until we heard shouting. And dogs. Someone angrily demanded to know what we were doing on his property. He had no flashlight, so we could not determine his location. Our headlamps were obvious, so he knew exactly where we were. My colleague shouted back details about his government-funded scientific research on the evolution of anole lizards. Importantly here, there was no language barrier because I was not doing the talking. This guy also had a shotgun. In response to my friend's eloquent description, he responded with "Bullshit. You're stealing my cattle!" and fired at us. I could hear pellets hitting leaves around me, and the dogs were getting closer as we ran, terrified and disoriented, through the forest hoping to clear the fence before he killed us. It is not easy to run through a cloud forest at night. In a shocking bit of hubris, the next morning we returned because this patch of forest was too valuable to be ignored. We introduced ourselves properly, gave a better explanation, and apologized profusely for trespassing the night before. He was really nice and let us camp on his property for a couple of days. Lessons duly learned.

When all goes well, and you can convince people that you are not insane or involved in subterfuge, local folks can be invaluably helpful and fun. While they may not know all the local animals, they do know the local trails and can give you great tips on the area. I often carry a deck of flashcards with pictures of the animals I am seeking. It's like a bad detective show: "Excuse me sir, have you seen this snake?" Kids especially will delight in helping you chase common lizards. If you give them a flashlight and a sack, they will deliver you a load of frogs each

Examples of herpetological flash cards used to communicate across language and cultural barriers during field work. Photo by Joseph R. Mendelson III.

morning. It is a well-known trick of tropical herpetologists to travel with coins to pay kids a bounty for their finds. Some kids are really good at this, and you can greatly improve your efforts with an army of them on the payroll. There always is a bit of financial bargaining involved, and the size of the creature, rather than the rarity or novelty, tends to drive the price. So, I have paid good money for loads of enormous Cane Toads that I did not need in order to scoop up the occasional "great find" that a kid brings in. One advantage about working with kids is that they don't care about why you want these animals. They are just having fun and earning some change.

Yet, miscommunications happen. I once watched a group of small boys struggling to haul a big burlap sack into my camp. The sack was bouncing crazily, loaded with a whole lot of something that was unhappy to be in there. I was suspicious, and when I asked the kids what was in there, they explained proudly how many rats they had for me. I re-explained to them that I wanted lizards and frogs and certainly not rats. I paid them a bit for their evident efforts and convinced them not to open the sack right there. I have no idea how they managed to catch so many rats. And I don't know if I failed to communicate my interests to them or if they were just trying to pull one over on me. Similarly, in Veracruz, Mexico, one morning, I awoke in the back of my truck with the sense that "someone was out there," so I emerged and found a guy I had met on the trail the day before. He was holding an enormous, gorgeous Keel-Billed Toucan. I thanked him and re-explained that I was looking for toads, not toucans, and encouraged him to release it. It flew off happily, he walked away nonplussed, and I have no idea how he managed to catch that bird.

Most people will not capture a live snake for you, but they love it when you will remove snakes for them. These people don't care why you want the snake and are just grateful you will take it away. This sounds like a great plan, until people rush into your camp in a frenzy, demanding that you follow them to collect this big snake they have found. You run after them, only to discover that the snake had been there days ago or even last year. To be polite, I'll look around for the long-disappeared snake and then apologize that I failed to help. Once in a while, however, there really is a snake. I followed a frantic group of people a long way from my camp to their village where a crowd had gathered at the base of a large palm tree. Pointing up excitedly, they indicated a large nonvenomous Tropical Chicken Snake in the tree. I wanted that snake, and they clearly wanted it gone. I cannot climb palm trees and, if any of these folks could do so, they were not admitting it. Not at all believing that my plan would work, I started hefting large fallen palm fronds up toward the snake to dislodge it. It worked and suddenly about 7 ft of angry serpent rained down on me. The crowd scattered, screaming. The snake landed on my shoulders and quickly took off into the grass. I managed to grab it, and it began biting me and thrashing around violently. I got a hand behind its head, so the biting stopped, but I was not controlling the thrashing until the snake suddenly wrapped itself around the length of my leg. I realized this was a blessing because it stopped thrashing. I made the long walk back to my camp, marching through the center of their village stooped over holding the head of a very large black and yellow snake covering my entire leg, looking like a barber pole. This made for quite a parade with a line of dancing kids waving those useful palm fronds and shouting. I was the freak at the center of the parade. This scene was so bizarre that I well recall thinking to myself at the time, *Why do I do this for a living?* The peoples' reactions were so joyful with hilarity that I thought to myself, maybe I do this because it is fun! I honestly think they were laughing too hard to even bother with wondering why I was doing this.

After a field season, it's rewarding to think about the amazing animals encountered and the new discoveries. But it's the people I encounter that leave the greatest impression on me. As often as I tell these stories to my friends, I am quite sure there are loads of people in Mexico and Central America who still are telling their side of the same story. *Once, a number of years ago, this fellow from the United States showed up claiming to be looking for snakes and toads. And it turns out he wasn't lying!* While my storyline about being a schoolteacher visiting from the United States seems to satisfy some people, most just don't grasp why I would put so much effort into chasing down weird or scary animals that are best avoided. Field encounters in the United States really aren't all that different. I realize that language and cultural barriers are not the issue. Most people just fail to understand why I do what I do. I'll admit that sometimes, in fits of frustration or discomfort in the field, I catch myself wondering why I am doing this.

My answer to myself is that I do this because these creatures fascinate me. I have gone for decades now obsessively wondering what might be under the next rock. But try as I might, I often fail to get that point across in any language. I don't think even my parents ever really understood this either, but they were relieved that I managed to get a paycheck out of being weird.

About the Author

Joseph R. Mendelson III received his PhD at the University of Kansas. Joe is director of research at Zoo Atlanta and adjunct professor of biology at Georgia Tech University, where he teaches regularly. He is past president of the Society for the Study of Amphibians and Reptiles. He coauthored the global International

Union for Conservation of Nature Amphibian Conservation Action Plan and cofounded the Amphibian Ark. Joe has studied herpetology for more than 30 years, concentrating mostly on Mesoamerica and the southwestern United States. Field and lab work led to the discovery and naming of about 40 new species. Other studies considered ecology, biomechanics, and natural history. Joe's essays have appeared in a wide variety of media, and he has published more than 125 technical papers in peer-reviewed journals.

42

THE CAPTAIN AND THE FROG

Sinlan Poo

I should begin by saying that although this may not be the typical field her-petologist story, it has woven in and out of my time in the field for more than 10 years. But let's start at the beginning, at my first field research experience in herpetology . . .

In the summer of 2006, I was a junior in college and had been given the oppor-tunity to spend a few months as a field assistant in Panama with my lab group from Boston University. It was a relatively easy decision: a free trip to Panama and a paid position working with one of the most charismatic frogs in the Western Hemisphere. That's how I got hooked in—by a young graduate student holding up a printed photo of a Red-Eyed Treefrog (*Agalychnis callidryas*) in front of the classroom and asking if anyone wanted to work with these colorful creatures. Apprehensions I had of joining a field team were quickly overcome by the excite-ment of exploring the Neotropics and finally seeing these frogs in their natural habitats. Our group was based in Gamboa, a small town on the banks of the Panama Canal that was built in the 1930s to accommodate canal workers. Gam-boa had a long history of hosting field biologists, though I was young and only vaguely aware of its historical importance in the pantheon of tropical amphibian research. I was focused on my first true taste of life in the field—the long, quiet days of walking from my colonial-style wooden house to the old school building that now served as our lab space, the slow visual searches around the edges of the pond to look for arboreal treefrog eggs that seemed like jeweled grapes, and the curtains of thunderstorms that periodically blanketed the town and forced everyone into a reprieve. It was a peaceful way of living, and I was enamored.

An avid and somewhat old-fashioned writer, I wrote long letters to friends back home in Taiwan and in the United States. A couple times a week, I walked around the back of the old school building and up the road to the small post office in town to send my letters and to pick up mail that had come for me in return. On the other side of the old school building was the Gamboa Union Church, built in the 1950s, shortly after the town was established. Not being a religious person, I walked by the church without much thought on most days. That is, until one day a few weeks into my summer.

I had just finished my daily duties of taking care of the Red-Eyed Treefrog eggs in the lab and was leaving the old school building when a tall, skinny Caucasian man walked up to me.

"Hi, I'm Pastor Bill from the Union Church next door," he said in an American accent. "I heard there was a Taiwanese student working with the frog researchers this summer, so I wanted to introduce myself and ask for a favor."

Being the only Asian around, I supposed I was rather easy to spot. Paster Bill was friendly and earnest—something that perhaps came with his profession—and we stood by the school building and entered into one of the most unexpected conversations I've had, before or since.

This favor, he said, was to accompany him to a prison a couple miles down the road for his Thursday morning service. He said there was an old Taiwanese cargo ship captain who was imprisoned in Panama for manslaughter because of some technicality and had not had any visitors since he was sentenced three years ago. "The Captain," as they called him, did not understand Spanish and only spoke limited English, so he had been isolated in more ways than one. I agreed but said that I had to make sure my work schedule allowed for it. Conveniently, one of the perks of working with Red-Eyed Treefrogs is that they are nocturnal and, consequently, so is one's work schedule. This meant that leaving for a couple hours in the morning, other than making me sleep-deprived, was not hard to manage. I should point out that I had never been to a prison, let alone a prison in the middle of a tropical forest in a foreign county. In the couple days that followed, during fieldwork, I wondered what The Captain had done to end up here. Paddling the small lab canoe around the pond and wading waist deep into the water to examine the development of the arboreal frog eggs, I wondered what I had gotten myself into, what I had agreed to, all too readily.

I was a new addition to the small group of retired European and American expats and missionaries that went through the prison gates that Thursday. The open-air prison looked like an old school building, much like the one I had just left earlier that morning, but with metal bars and meshed fences for windows. To the side, we entered a chapel-like gathering hall, and I spotted The Captain amidst the crowd of Panamanian inmates immediately, as he must have spotted me amongst the Caucasian visitors. He looked to be in his sixties and reminded me of my grandfathers—upright in posture due to his military background, calm and soft spoken, reserved, and extremely polite to this 21-year-old he had just met—which was especially surprising considering how elders are respected in Chinese culture and how young I must have seemed.

Not knowing what to say, I told him about the frog research I was doing, about the Red-Eyed Treefrog eggs we were collecting in the ponds nearby, and the predator simulations we were doing to determine what was triggering the escape response of hatchlings from the egg capsules. He had never been outside the prison walls, let alone into the field in Panama, but I told him about the ponds that were just a few miles up the road, and he talked about the frogs he'd seen in the fields as a kid. He seemed genuinely interested in hearing about "my frogs." Perhaps it was an escape or simply a conversation in one's mother tongue with someone from home, but the frogs became our touchstone.

From then until the end of my summer internship, I visited The Captain every week. We only ever had three to five minutes to talk during each visit, which was the amount of time between when we walked into the chapel and when everyone was seated and ready for Pastor Bill's service to begin. I gave him weekly updates on my fieldwork, while every now and then he would share a small anecdote from his childhood or youth in the navy.

Not long after, the summer ended, and I returned to the United States to complete my undergraduate studies. I started writing letters to The Captain that I would mail to the Gamboa Union Church, which Paster Bill would then pass along to The Captain. And, once or twice a year, Pastor Bill would mail me a letter back from The Captain. The letters would take weeks or months to go from my hands to his and from his to mine. To my surprise, the first correspondence I got from him contained a worn newspaper clip of an article about a frog. I'm sorry to say I don't remember what the story was about. What I do remember was reading his neat, traditional calligraphy-like handwriting saying that Pastor Bill had brought him a Chinese newspaper from a local fruit stand and that he thought I might find it of interest. How long he had kept that newspaper clipping I am not sure. I could see that the edges were torn by hand through folding the paper. The simple fact that he thought of me while looking at a newspaper story about frogs somehow conveyed a deeper emotion that, even now, is hard to put into words.

The next summer, I graduated from college and moved to Southern California to work as a field herpetologist, doing baseline surveys of endangered amphibians and reptiles under the arid climates of the Inland Empire. I wrote to The Captain about my new job, the new habitats I was surveying, and the new frogs I was seeing in the riparian habitats and vernal pools. And in return, he sent me letters of encouragement and assured me of his good health. As I got into graduate school,

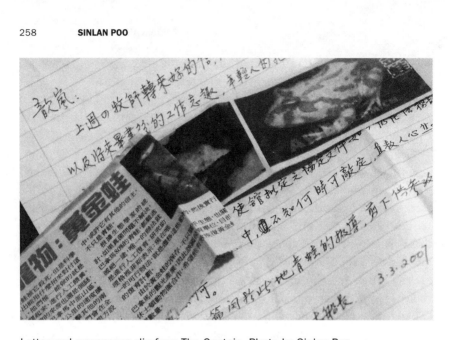

Letter and newspaper clip from The Captain. Photo by Sinlan Poo.

my pursuits took me to the tropical forest and ponds of Thailand, and I looked forward to sharing with The Captain the new frogs I was studying. However, I found out that, after years of requests, The Captain had been granted clemency and had been extradited to a prison in Taiwan. This was wonderful news, yet it also meant the end to my roundabout way of maintaining some correspondence with him. That was it.

Years passed, and as I waded chest deep in a different type of tropical pond looking for a different type of arboreal frog eggs, I wondered how The Captain was doing. After I finished my doctorate in Southeast Asia, I moved on to my postdoc in Tennessee and started working with a new set of amphibians in yet another new ecosystem, until one day I learned that the research project I was working on would once again take me back to the field station in Gamboa.

With this came a sudden urge to share this news with The Captain, though it had been years since we lost touch. I searched through newspaper clips and did some amateur sleuthing online. As luck would have it, he had a rather unique two-part surname, and I wondered how many people would share this surname in Taiwan. Turns out, there was only one on Facebook. It wasn't him—not that I expected it to be—but about a day after I sent a very tentative message saying, "I'm sorry to bother you. . . . This might sound far-fetched, but do you happen to know . . ." I got a response "Yes, he is my father. Are you the researcher who visited him in Panama? He told us about your work with frogs."

A few months later, I was sitting at a Häagen-Dazs store at the Taipei train station talking to The Captain about "my frogs" again. He had been out of prison for a couple of years and was now living with his family and taking care of his grandchildren. It was surreal, to say the least, though I'm not even sure if surreal is the right word to use. We seemed worlds apart from when we last sat together, and yet here we were again, him asking about my research and taking pleasure in hearing my description of the field, the ponds, and the frogs. A few months after that, I arrived in front of the old school building in Gamboa. I caught up with Paster Bill, who had stopped going to the prison because of a recent bout of illness but was still holding his regular Sunday services. I showed him a photo of me and The Captain at the ice cream shop on my phone. It had been 11 years since that summer.

Walking around the old ponds, seeing the translucent egg clutches hanging over the water, and once again admiring the beauty of the Red-Eyed Treefrogs, I considered how a printed photo of this frog had turned into a career for me. There is something about the calmness of the field, the isolation, and the peace it brings that allows for reflection . . . like the evaporation of noise that allows one's thoughts to distill into clear droplets of water. In some way, my first field experience with the Red-Eyed Treefrogs has influenced everything that has followed. But beyond the academics of it all, it is the connection with The Captain that I think back on most frequently and perhaps one of the experiences that I cherish the most—the connection through these small creatures that first broke the ice and, I would like to think, offered a mental escape for him and showed me how easily they connect us all.

About the Author

Sinlan Poo (PhD, National University of Singapore; BA, Boston University *summa cum laude* with distinction) is the curator of research at the Memphis Zoo with an adjunct position at Arkansas State University. Originally from Taiwan, she is a behavioral ecologist and conservation biologist with broad interest in parental care, reproductive ecology, and linking in situ and ex situ conservation. Dr. Poo's research is primarily focused on amphibians but has included a wide range of projects on reptiles, small mammals, carnivores, invertebrates, and rare plants across several countries (the United States, Ecuador, Panama, Singapore, Sri Lanka, Thailand, and Taiwan).

43

EXPLORING THE WILD KINGDOM WITH MARLIN

Lee A. Fitzgerald

Just as the director of the National Forest Service of Paraguay opened the door to our office, the room went dark as an 8-in rubber band smacked the light switch next to his head. Even though it was a great shot we stopped celebrating. "*¡¡¿Que pasa aqui?!? ¡Este oficina es un lugar de trabajo!*" ("What's happening in here? This office is a place of work!") Well, for us this was important work! When your shot turned the lights on or off, you scored. We had devoted the entire day to this task as Peace Corps volunteers (PCVs) in Paraguay. After all, we could not be efficient field biologists, charged with discovering and documenting the herps of Paraguay, without honing our expertise in the rubber band lizard-collecting technique. After trying to explain this to the director, Ingeniero Hilario Moreno, he called us into a meeting.

From 1979 to 1982 I was a Peace Corps volunteer. I spent the first year in El Salvador, where my job was to study iguanas and ctenosaurs at Laguna El Jocotal National Park alongside local community members who had become park rangers. I learned Spanish, lived for nine months without electricity and not even an outhouse, became *compañeros* with my local counterparts, and had an absolute blast marking hundreds of iguanas and ctenosaurs with this band of rangers who wielded their machetes with the precision and ferocity of samurai swordsmen. As war broke out, I witnessed acts of terrorism, bombings, and machete fights, and we found murdered corpses when patrolling the park. It was a profound and emotional year for me. I happened to have malaria when I was one of the last to be evacuated from El Salvador a couple of weeks before Archbishop Romero was assassinated on the steps of the cathedral in San Salvador on 24 March 1980.

I boarded a plane for the second time in my life and headed straight to Paraguay via Miami, where I bought a boom box in the airport. I had been given a unique opportunity that would shape my entire career as a biologist. A couple of PCVs in Paraguay, David and Diane Wood, had conceived the idea that if conservation and environmental education were to make any progress in Paraguay, the country needed a national biological inventory. Paraguay was essentially unstudied by twentieth-century herpetologists. Diane and David arranged partnerships between the government of Paraguay, Peace Corps, US National Museum of Natural History (NMNH), and US Fish and Wildlife Service (USFWS) Office of International Affairs. The overarching goal of the project was to create the first Museo Nacional de Historia Natural de Paraguay (National Museum of Natural History of Paraguay).

With the National Biological Inventory in place, the stage was set for the initial team of PCVs to work with National Forest Service professionals, Paraguayan university students, and scientists from NMNH and USFWS. These leaders included Drs. Mercedes Foster (Ornithology), Norman J. Scott Jr. (Herpetology), Gary Hartshorn (Botany), Don Wilson (Mammalogy), and Paul Spangler (Entomology). Looking back, I can hardly believe my incredibly good fortune suddenly to be working with some of the most prestigious tropical biologists of the time. Our Paraguayan counterparts included a gifted group of young students, mostly women, including Aida Luz "Lucy" Aquino, Nancy Lopez, Lidia Perez, and Isabel Gamarra, who all went on to staff and run the Museo Nacional and have great careers. Lucy and I were mentored by Norm, and this is when I became a herpetologist. We have remained close friends and partners in Paraguayan herpetology since our first expedition to Ybycuí National Park in March 1980. From that point, my purpose in life was to get into the field as much as possible, venturing to remote places and collecting specimens for the yet-to-be museum. We were consumed with discovery, natural history, and identifying what we found. We thought this was essential for any kind of conservation to take place. How could you do conservation and environmental education if you had no idea what species were in Paraguay? How could we integrate the rich Indigenous knowledge, including all the Guaraní names for species? After Latin and Greek, Guaraní is the third most common language in scientific nomenclature, and many species had Guaraní names long before Linnaean names. A new national museum would be the center of it all. I've had many adventures, but these collecting expeditions in Paraguay will always be the most sublime and enjoyable fieldwork of my life.

The National Biological Inventory was created before the term biodiversity was coined and before Paraguay even considered research and collecting permits. It was the first effort at organized scientific collecting expeditions since the Paraguay Expedition of 1858 and the contributions of Moises Santiago Bertoni

(1857–1929). Paraguay, and in particular the Gran Chaco, the expansive biome of dry tropical forest that covers an environmental gradient from seasonally flooded wetlands in the east to very dry thornscrub farther west, were very undeveloped in 1980. The Paraguayan Chaco was not deforested, rarely visited by outsiders, and full of wildlife. We regularly interacted with Chacoan peccaries and the two other peccary species, saw Brazilian tapir, anteaters, and pumas regularly, and discovered so many herps that we could not identify. We found Budgett's Frogs (*Lepidobatrachus* spp.) and Waxy Monkey Treefrogs (*Phyllomedusa sauvagii*) and had no idea what they were despite having all the literature Norm could find. Now they are in the pet trade. For us, Paraguay was an open book, and we got to fill in some of the pages. This National Biological Inventory project remains a great success story for the Peace Corps and for the natural sciences in Paraguay. Today Paraguay's National Museum of Natural History is doing great with its third generation of curators. Paraguay has its own herpetological society, the Asociación Paraguaya de Herpetología.

Back at that meeting that interrupted our busy workday, Ingeniero Moreno informed us that the Servicio Nacional Forestal and Peace Corps had been contacted by *Mutual of Omaha's Wild Kingdom* TV show, starring Marlin Perkins. This was a big deal, since Wild Kingdom was the longest-running prime-time show on TV from the 1960s through the1980s, and Marlin was a biologist known throughout homes in America. The producer wanted to do an episode featuring PCVs and had gotten wind of the National Biological Inventory. I immediately thought it would be cool if we could demonstrate our skills with rubber bands. The episode would also include Dr. Mercedes Foster, curator of ornithology at the NMNH, who had been deeply involved in the project. Mercedes, a classically trained ornithologist and tropical biologist, was also told by her bosses at USFWS that she would need to use a net-gun in the show (more on that later).

I had no idea how TV shows are made, and I still laugh about it. And cringe about it. The first part was awesome. Suddenly we got to go to the field as much as we wanted to scout locations and catch animals that would be in the show. The setting for the film would be the Gran Chaco. We took the Peace Corps' pickup to a fantastic ranch, La Golondrina, where Diane and Antonio Espinoza hosted us, and spent days catching herps that we might use in the film. One night I found a Yellow Anaconda (*Eunectes notaeus*) eating a Wood Stork. We caught Yacaré Caimans (*Caiman yacare*), 17 species of frogs, and too many other cool snakes to list except for one that cannot go unmentioned—the large, feisty, and hard-biting False Water Cobra (*Hydrodynastes gigas*). I was always dirty and wore torn T-shirts, short jogging shorts, and Army jungle boots with no socks.

Filming began when Donald Meier Productions sent their cinematographer, Peter Drowne, to build the story line and get all the wildlife shots. I was named as the helper and would be one of the PCVs featured in the show. The first thing that distressed me was Peter and I went to the market, and he bought three identical sets of nice khaki pants and shirts for me to wear at all times while filming. This was not cool. I never dressed that way, especially when working! Once I got over the wardrobe, we had a lot of fun, spending days and days in the field building blinds and waiting for wildlife and using his 16 mm film camera with big reels of film and big old-fashioned batteries and beautiful lenses. I watched him film gray brocket deer eating flowers out of a pond as they fell from a bottle tree. He filmed lots of bird life, such as Jabiru Storks, Maguari Storks, Wood Storks, egrets and herons, and Neotropic Cormorants. The cormorants had converged on the drying ponds to feast on the armored catfish (*Hoplosternon* spp.). With these shots, Peter developed his storyboard that, of course, had to include the net-gun.

We spent days figuring out what to do about the net-gun. Mercedes had arrived to do her part handling this monstrosity. It is a heavy tripod bolted onto a rifle stock. The legs of the tripod, pointing out, each hold a football-shaped projectile that is attached to a triangular net. When you pull the trigger, three .22 caliber blanks inside the foam footballs shoot the net out, which is supposed to catch a bird on the wing unharmed. Mercedes was to walk around the Chaco being filmed catching birds this way! Great idea, Fish and Wildlife Service!

Thanks to Mercedes, my first publication was "A technique for live-trapping cormorants," which has absolutely nothing to do with the net-gun. Stymied by the net-gun, Mercedes suggested we make modified bal-chatri traps with mono-filament nooses tied to a flat piece of screen staked to the ground. We did and laughed our butts off as cormorants would waddle up the bank of the pond to sun their wings and get their feet stuck in the nooses. They would just stand there, like, "Weird, my feet won't move." Then we would leave our blind and retrieve a couple. If you ever watch Wild Kingdom's "The Unexplored Gran Chaco," you will see Dr. Mercedes Foster and Paraguayan ornithologist Nancy Lopez—guess what—walking around the Chaco catching cormorants with the net-gun. I still think walking around shooting rubber bands at lizards would have been more authentic.

Marlin Perkins was not just the star of an Emmy Award-winning show, he was an excellent and very well-respected herpetologist. He had been the curator at the Chicago Zoo and then the St. Louis Zoo, where he influenced his field and inspired many people. When I met him in 1981, he was 78 years old and vibrant. The Peace Corps directors held a reception for him at their home in Asunción where we all went to meet him. Marlin was a great storyteller and gentleman, noted for his high-pitched charismatic voice. While a group of us were surrounding him, he said, "Let me tell you about the time in 1929 when I survived the bite of a Gaboon Viper!" (with emphasis on "GabOOON VIPER!"). Wow, what a story. He told us they couldn't detect a pulse after he was bitten, but he lived. We introduced ourselves, and he said, "So Lee, are you publishing in herpetology?" Me, exhibiting FIS (full imposter syndrome): "No, But I want to! Uhhhh, I'm getting pretty good with the rubber bands!"

I spent more than a week with Marlin, going to La Golondrina to get some shots. All of this was great fun, and hanging out with Marlin was a hoot. While filming with Marlin, we would walk around posing by trees and bushes to get shots that would be edited into the wildlife shots. These are called look-see, where they film you looking for wildlife, then the scene cuts to the brocket deer eating flowers weeks before and you act like you had just seen it. We used a snake I found and kept in a pillowcase for a rousing scene where Marlin and I chanced upon and wangled this impressive 7-ft plus False Water Cobra. Marlin loved doing that, and he knew what he was doing. I got to fly in a Cessna airplane with Marlin because we had to get some aerial shots for the film, which included me "checking on some projects" with the National Biological Inventory, sporting

Marlin Perkins and Lee Fitzgerald holding a Yellow Anaconda in the Paraguayan Chaco in 1981. The photo was taken during the filming of an episode of *Mutual of Omaha's Wild Kingdom* entitled, "The Unexplored Gran Chaco." Photo by Peter Drowne. Reprinted with permission of Mutual of Omaha Insurance Company.

my khaki shirt and oversized walkie-talkie. We still howl about that. Getting the shot of us flying over the Chaco preceded *Top Gun*. At the airport, Marlin and I situated ourselves in the plane. Peter set up his camera on an angle and focused on the cockpit while an assistant shook the wings of the plane to make it look like we were flying and pointing out features on the Chacoan landscape below. That's acting!

It turns out they know to film these things only on sunny, cloudless days, so that all the shots fit together as if it were one flawless fieldtrip. There was a lot of down time at an obscure locality, Agua Dulce near Cerro León, while waiting for a cloudless day. A group of national park rangers and PCVs was along to participate. It was hot as blazes. We were sitting in our underwear playing poker in the heat of the day when Marlin walked out of the rustic cabin in full-blown Wild Kingdom attire, neatly pressed and looking sharp. In characteristic Marlin inflection: "Hello boys! I was wondering if you'd like to take a break from that poker game and hear about my search for the Yeti!, in the Himalayas!, with Sir Edmund Hillary?!?" Us, immediately putting down our cards, "Yes Marlin! Tell us all about it!" Wow again, what another story!

A couple of days later it was over. Marlin was headed directly to Mexico to do another episode. When we were saying goodbye, he asked in his characteristic voice, "It's been great getting to know you, Lee. I'm off next week to go scuba diving with hammerhead sharks in Baja California! Is there anything I can do for you?"

I said, "Marlin, I haven't seen my mother in two and a half years and only talked to her by phone twice during all this time. Could you give her a call for me when you get back to the United States?"

"Why of course, Lee!" A few weeks later, after Marlin did the show on hammerhead sharks, he surprised Mom with a call. "Hello Mrs. Fitzgerald, this is Marlin Perkins from Mutual of Omaha's Wild Kingdom! If you have time, I'd be delighted to talk to you about spending time with your son Lee in Paraguay!"

About the Author

Lee A. Fitzgerald, professor and curator, Texas A&M University, is a herpetologist, conservation biologist, and ecologist. Lee's research has focused on sustainable use, wildlife trade, and mechanisms determining persistence and extinction of species. He has advised 19 PhD and 20 MS students and taught herpetology 25 times to 824 students. He directs the Applied Biodiversity Science Program and is chair of the Ecology and Evolutionary Biology Program at Texas

A&M and has served herpetological societies in various capacities. His honors include Outstanding Peace Corps Service, Sigma Xi Excellent Dissertation 1993, Texas A&M Vice Chancellor's Awards for Undergraduate Teaching, Graduate Teaching, International Programs, Interdisciplinary Team Research, and Alfred P. Sloan Foundation Minority PhD Mentor. Lee has published more than 185 scholarly works on lizards, snakes, turtles, crocodilians, frogs, birds, elephants, and people.

44

TERROR, COURAGE, AND THE LITTLE RED SNAKE

Tiffany M. Doan

In May 2011, I set out to teach my biannual tropical ecology college class in the Peruvian rainforest, a course I had been leading for the past eight years. I took undergraduate and master's students from Central Connecticut State University to the Amazonian lowlands to encounter the amazing biodiversity and to teach the methods of field ecology and the scientific process. Students designed and completed small original research projects on a variety of organismal groups, which included (depending on the year) mammals, birds, insects, plants, and reptiles and amphibians. In addition to experiencing the wonders of the rainforest, students were immersed in Peruvian culture, working with Peruvian scientists for their projects and living at a functioning research station with Peruvian workers and scholars.

For that year, we traveled to a remote field station known as Sachavacayoc Centre, in the Tambopata Province of southeastern Peru, which is on the western edge of the Amazon Basin. Sachavacayoc Centre was one of my favorite localities for this class because it is primarily a research station as opposed to many of the other stations in the region, which were mainly tourist facilities but also hosted researchers. The nearly pristine lowland rainforest at the site has an incredible diversity of plants and animals. For most of our visit, we were the only researchers there, and we felt like it was our own private oasis in the jungle. The accommodations were rustic, to say the least, with no electricity and brown drinking water that was pumped up from the river. But we had roofs over our heads, beds, mosquito nets, a basic laboratory, and a kitchen staff that prepared delicious local food. As there were no roads, we arrived by boat after a three-hour trip from the rainforest town of Puerto Maldonado. My students felt like they were nineteenth-century explorers experiencing uncharted territory and a new way of life for the first time.

As a herpetologist, I loved to expose my often all-too-sheltered students from Connecticut to exotic, scaly creatures such as lizards, snakes, and caimans. One of the things that has always fascinated me about the students—indeed, about people in general—is that so many are terrified of snakes, which research has documented as an innate fear shared by humans around the world. A common misconception I encounter is that all snakes are venomous. In fact, most snakes in the world do not produce venom. In South America, venomous snakes are quite uncommon, and one needs some luck to encounter one. Most of my students had never seen a snake in the wild, and if they had, it was likely a garter snake slithering quickly by in the woods near their New England homes.

Despite people's trepidations about snakes, serpents also inspire fascination, as reflected by the sheer number of nature documentaries, sensational news stories, and horror movies that have been produced about snakes. During my 18-day class that summer, I searched tirelessly to find as many snakes as possible to show to the students. You might call me the anti–Indiana Jones. That year, my course was divided into students who studied (1) birds, (2) mammals, and (3) reptiles and amphibians, or herpetofauna. My "herp" students were more-or-less fearless, flipping over logs, crashing through the dense vegetation, and exploring oxbow lakes to find as many frogs, lizards, caimans, turtles, and snakes as they could. I had taught the herp students how to capture frogs, lizards, and snakes, taking care to disturb them as little as possible. Because of the seven species of highly venomous snakes found in the area (out of over 60 snake species), including bushmasters and coral snakes, my rule for the students was that no one was allowed to catch a snake unless they could recite its scientific name, or if the snake was primarily red. None of the five species of red snakes in the region is the slightest bit venomous or aggressive.

Late one night, while we were surveying for frogs, Andy, one of my graduate students, encountered a tiny red snake, *Drepanoides anomalus,* commonly called the Amazon Egg-Eater Snake or the Black-Collared Snake. He asked permission to capture it and gently lifted it from the ground. This relatively common species, which ranges throughout six South American countries, is about the least-threatening snake one could encounter, measuring up to only 50 cm and eating only lizard eggs. These snakes often hide their heads within their coils, as an ostrich buries its head in the sand, to protect themselves. I realized this would be the perfect snake for the mammal and bird group students to see, and hold, if they gathered the courage. We deposited the snake in a cloth bag for its journey back to the lodge and pulled it out the next morning after breakfast to show the students of the various teams and the local dining room workers.

Herpetology students in the Peruvian rainforest in 2007 after the capture of a Rainbow Boa (*Epicrates cenchria*). Photo by Tiffany Doan.

As I taught the students a little about the snake's biology, several students worked up the courage to touch the middle of the body of the snake. A few even volunteered to hold it. Like so many frightened people before them, the students were surprised by the smooth, nonslimy texture and enjoyed watching the snake tongue-flick to smell its surroundings.

Nicole, a 22-year-old petite undergraduate student studying mammals, who had always been driven to terror in the presence of snakes, decided that, right then and there, she would conquer her phobia and hold the snake. She held out her hand, and I gently placed the calm, coiled snake in it. Despite the snake being in her hand, she kept her eyes trained across the room out of the window, her head cocked back at over 90 degrees. It took me a few moments before I realized why she peered over her shoulder—Nicole could hold the snake as long as she did not see herself holding it. The tiny snake cooperated beautifully and did not move the entire time it was in her hand. Their mutual fear kept both the snake and Nicole very still.

After over a minute of holding the snake, I removed it from the student's hand. Nicole was overjoyed. She told me that she had conquered her fear of snakes. She had accomplished something that she never thought she could, overcoming the fright that had been with her since childhood. She later remarked that the experience of holding the snake was one of the best moments of her trip—even more memorable than watching howler monkeys frolic, observing capybaras munching on river vegetation, or exploring the Incan ruins of Machu Picchu.

I've been teaching international biology courses for over 20 years at this point, helping students grow as scientists and as global citizens. After all this experience, I knew that my trips to the Amazon were important adventures in the students' lives, but I never expected that one minute with a little red snake would be so transformative.

About the Author

Tiffany M. Doan is an associate research biologist at Archbold Biological Station. She received a BS in biology from the University of Miami and a PhD in quantitative biology from the University of Texas at Arlington. She studies the evolutionary biology, ecology, and conservation of reptiles and amphibians, specializing in the lizard family Gymnophthalmidae from the Andes and Amazon of South America. She has described six lizard and one snake species as new to science, studied how malaria affects lizard species, examined species distribution patterns of reptiles, and works to conserve endangered reptile populations. She is an avid scuba diver and aikido martial arts practitioner.

45

TEAM SNAKE MEETS
EQUIPE SERPENT

Kate Jackson

When I was a doctoral student, I started doing herpetological fieldwork in the Republic of Congo. When I was a postdoc, I started taking Congolese undergraduate students with me in the field because, like biology students everywhere, they need research experience and that was something I could provide. Some of the same students continued to work with me over many years, and we have grown up together as scientists. Ange Zassi-Boulou completed his doctorate a couple of years ago, and Lise Mavoungou is close to finishing hers, with me as cosupervisor. We are now colleagues and collaborators and call our Congo research group "*Equipe Serpent.*" I have been a professor at Whitman College, a small liberal arts college in Washington State, for the past 14 years, and I and my research students there call ourselves "Team Snake."

Here I am writing about an expedition to Guinea in July 2021. This expedition was particularly important for me because it was the first time that *Equipe Serpent* and Team Snake were together in the field. The reason was that Jordan Benjamin, one of my former Whitman College students, had founded the nonprofit Asclepius Snakebite Foundation to tackle the problem of snakebite in sub-Saharan Africa. He runs it with his friend Nick Brandehoff, an American physician. In Guinea, a clinic run by herpetologist Cellou Baldé has been doing a pretty effective job of treating snakebites under very austere conditions. With more resources and training, the clinic could be even more effective, and this is what Asclepius aims to provide. Meanwhile, in Congo, there is no effective snakebite treatment available. Jordan, Nick, and I went to Guinea, and Lise and Ange joined us there, to get a sense of what effective snakebite treatment looks like in a country not very different from their own, so that we could work toward making snakebite treatment available in Congo. We were hoping that Congo could learn from Guinea and that maybe there would be things Guinea could learn from Congo. We also did lots of herping, alongside Guinean graduate student herpetologists Alpha Baldé and Martin Millimouno.

July of 2021 was a welcome moment of calm in the COVID-19 pandemic that had upended life in the United States for the past year. Vaccinations were available, COVID rates were down, and travel was once again possible. There we were, against all odds, in Guinea. The research institute in Kindia is a large complex of colonial-era buildings, including one for "venimologie," housing the herpetology collection (fluid preserved), the live snake collection, and lab space, and one small building serving as a snakebite clinic. There's also a guest house where we stayed, right on site. The campus is surrounded by wonderful herp habitat—rocky open areas with flowing water, cliffs and outcrops, tropical forest, villages, and small-scale crops.

Something I had not foreseen was how tiring it could be to be called on without warning to translate. I am Canadian and can function effectively in either French or English. But I must normally spend a lot of time lost in my own thoughts and not paying much attention to the conversations going on around me, so when someone suddenly said, "Hey Kate can you ask Alpha if . . ." or "Kate, *s'il te plait, explique à Jordan que . . .*," again and again I had to explain, "Sorry, my head is full of snakes," "*J'ai la tête pleine de serpents.*"

We found our way together as a group gradually. Jordan is a wilderness medic as well as a herpetologist. Like me, he can immerse himself completely—lose all sense of time and place—when focused on something exciting involving venomous snakes. Unlike me, he seems to experience the same sort of thrill helping people in medical emergencies, and he's very good at it. On the first day, while Alpha and Martin and I went out looking for snakes, Jordan worked in the snakebite clinic with Nick. "Don't forget to take water with you and to drink it," he reminded me, "otherwise you're going to get an IV." We found no snakes on that first day, and Alpha, as the local expert, must have been feeling stressed about that. But I've been a herpetologist long enough to know that hours and days of beating the bushes and finding nothing is more the rule than the exception, even if the highlights we remember and the stories we tell are mostly about those exceptions. You have to just enjoy a beautiful day in the field in an amazing new place. I tried to reassure Alpha as we headed back to the serpentarium at the end of the day. "The snakes can't hide from us for three weeks," I told him, with only slightly more confidence than I felt.

The next day, we were joined by an older man, Naby Keita. Long ago he was bitten by a mamba and brought to Kindia for treatment. Cellou saved his life.

When he recovered, he told Cellou he would never leave him. He now works as Cellou's snake catcher and has named one of his children Cellou. Together we tried a different place, once again just a short walk from behind the serpentarium, working our way upward over a rocky slope with water trickling down it in places. Soon we were in dense forest habitat, pushing our way through tangled branches and lianas, pausing to catch our breath and sift through leaf litter with our snake hooks. Within the first 30 minutes sweat had completely soaked my clothes and trickled into my eyes. We saw a few small skinks, but no snakes, and little chance of catching anything we saw because of the difficult terrain.

"Jungle habitat," said Alpha gloomily.

"Just about all our fieldwork in Congo is like this," I told him. "We do a lot with pitfall traps."

Alpha looked up and locked eyes with me. "What are pitfall traps?" I suddenly knew what the Congo herp team could contribute.

Ange and Lise arrived a few days later from Congo. On their first morning in Kindia, Lise decided to get started right away on the epidemiology work that she needed for her dissertation and headed to the snakebite clinic for the day, with Jordan and Nick. Ange opted to join Alpha, Martin, Naby, and me in the field, continuing our search for snakes. Again we spread ourselves out to cover the terrain. For the next several hours we searched trees, bushes, and other refugia for snakes, staying within a few yards or at least within sight of one another, mostly in silence and with all of us in private bubbles of focus on our own snake hooks and the rocks and trees and bushes nearby. The first time the bubbles burst was a false alarm. Martin pointed: "I thought it was a snake, but now that I keep looking, and it isn't moving, I'm thinking it's just a dead twig. If you look where those two big branches split and then out along the smaller branch . . ."

"Oh yes, I see it. *Thelotornis*, right?"

"Yes, exactly!" We laughed. It felt both validating and faintly ridiculous that the dead twig looked to both of us like the same snake, *Thelotornis kirtlandii* (Vine Snake).

But the next shout from Martin was not a false alarm. It was a Green Mamba, and it had gone up a tree—a perfect tree from which a team of herpetologists could catch a Mamba as long as we all got there quickly enough to surround it while Martin kept it in sight. "*Un serpent!*" Martin shouted with increasing urgency, "*un mamba!*" The tree was a short, stunted one growing out of the rocky ground, with another similar tree beside it but no other trees for some distance. Martin pointed where the snake had gone, and Alpha leapt up the tree after it, with Naby handing up the big tongs to him. The Mamba, a muscular streak of green and gold, glided up and out of reach along a branch and into the upper branches of the second tree, but Ange and I had both sides of that tree covered, keeping the snake from continuing down the trunk and onto the ground. Naby leapt up the trunk of the second tree and we passed the tongs up to him. Soon he had the snake caught midbody in the tongs and pulled out of the tree to an open rocky area where the rest of us could help maneuver it into a snake bag.

We all went out herping after dark when astonishing numbers of giant land snails and giant millipedes were out. "It must be their mating season or something," said Jordan. We scanned the trees with headlamps and flashlights, hoping for snakes, but excited by the sleeping chameleons we found this way. We returned to the guest house, tired and dirty and content. Jordan and Nick decided to check on the patients in the clinic before turning in for the night, and Lise and Ange went along with them, so only I turned in. I was roused from my room a couple of hours later by excited shouts of the others calling for me. They thrust a plastic bucket toward me, all talking at once. The bucket had been standing just outside the front door of the clinic, and luckily none of the patients or staff had noticed that there was a snake in it. The snake was glossy and mostly black and small enough not to be able to get itself out of the bucket after falling into it. None of them was sure what kind of snake it was, but they were all experienced enough herpetologists to know of a few small, glossy-black, highly venomous possibilities that they weren't certain it wasn't. Hence the decision simply to bring the whole bucket to me. In the beam of someone's headlamp, I looked at the little snake, glossy and black with a line of large bright-red spots down the middle of its back. It was a west African species that I had never seen before. But I'd seen plenty of central African ones in Congo and recognized the genus by the distinctive shape of its head as a (harmless) wolf snake (*Lycophidion*).

Over breakfast I noticed that Ange had his notebook open in front of him and was showing Alpha diagrams of pitfall trap arrays. I was so glad to see that he was explaining how to build a pitfall trap line so that the Guinean team would know how to do that, and I told him so.

"Oh no, we're planning to do much more than that," he replied.

Lise and Ange were planning to build a drift fence and pitfall trap array for and with the Guinean team and to monitor it with them for the first few days so that they really got the hang of it. "And also funnel traps for catching snakes," Lise added. Ange went to the market with Alpha after breakfast, and they returned loaded with plastic sheeting, wire mesh, plastic buckets, zippers, scissors, several staple guns, and lots of staples. The herpetology team got to work. We unrolled the plastic sheeting and measured and cut long strips of it for the fence. We punched holes in the bottoms of the buckets to let water drain out. The most time-consuming activity was making the funnel traps. These are like lobster pots and built out of fine wire mesh, cylindrical in shape and with a funnel-shaped entrance at each end. Their cleverest feature is the use of a zipper along the length of the cylinder so you

Building a pitfall trap array in the forest near Kindia, Guinea, July 2021. In the foreground: Martin Millimouno; middle (left to right): Lise Mavoungou, Ange Zassi-Boulou, Kate Jackson; in the background: Alpha Baldé. Photo courtesy of Martin Millimouno.

can remove a trapped snake and close it up again. The whole thing is held together with lots and lots of staples. By the end of the day we had a neat pile of all the materials needed for two pitfall trap arrays. The next morning, we set out into the forest to set up trap arrays in two locations suggested by Naby. Both arrays would have the same total length of drift fence and number of buckets, but one would be arranged in a single line parallel to a small stream at the base of an overhanging cliff, while the other would be three shorter fences radiating outward like spokes from the center, in a forest location without any streams or cliffs to work around. We spent most of the morning building the first one, and for most of that time we were on our hands and knees in the leaf litter, making sure that the bottom edge of the plastic sheeting was buried in dirt and camouflaged with leaves.

Over the next few days we caught small snakes, lizards, frogs, and even a small fish in the pitfall trap/funnel trap arrays. Several of these were species that the Guinean team had not seen before.

"I see that this is a *Psammophis* [Sand Snake], but this is a species we have not seen before," said Alpha in tones of astonishment, seeing a field site whose herpetofauna he knows better than anyone through the new lens of the pitfall trap array.

Lise was unsurprised by his surprise, "You may not always get more species using the pitfall traps than by active searching," she told him, "but you'll usually get different species."

As we made our way back at the end of the day, laden with specimen bags, we were spectacularly dirty—completely coated with a layer of forest-floor dust—from a day spent sweating in a wet tropical forest while crawling around and digging in the leaf litter.

"How are you holding up, Martin?" I asked, turning my attention to the youngest member of the team.

"It's been a good day," he twinkled, "*J'ai la tête pleine de serpents.*"

About the Author

Kate Jackson (PhD Harvard University; Honors BS, MS University of Toronto) is professor of biology at Whitman College, in Walla Walla, WA, and director of herpetology for the Asclepius Snakebite Foundation. Kate is a herpetologist whose research explores the morphology, biodiversity, and evolution of amphibians and reptiles, with a regional specialization in central Africa and a taxonomic focus on snakes, the snake venom-delivery system, and snakebite. Kate is the author of three books, including most recently *Snakes of Central and Western Africa* (2019, Johns Hopkins University Press, with coauthor J-P. Chippaux), as well as more than 40 scientific articles and book chapters.

46

TICKS, POLICEMEN, AND MOTHERHOOD

Experiences in the Dry Chaco of Argentina

María Gabriela Perotti

As a young biologist I had the privilege to work with herpetologist and friend Lee Fitzgerald, who taught me to love herpetology and fieldwork. Along with my husband Felix Cruz, a lizard herpetologist, we experienced many fun and some not so fun experiences in the field. While developing my doctoral thesis, our work was centered in Grand Chaco, in particular the dry Chaco region in Salta, Argentina. Our field sites were near the small town of Joaquin V. Gonzalez, in the center-west of Salta Province. During those days, in the 1980s, piercing and tattoos were not very popular in Argentina, especially not in the Salta forest. So imagine my surprise when I experienced a piercing experience: I discovered on my nipples and behind my ears huge ticks embedded in my skin. Waves and waves of ticks invaded our boots looking for interesting corners of our body. These were my first experiences in the dry Chaco, an adverse place but at the same time, a beautiful, amazing, and interesting region to study animals, especially amphibians, with so many biological peculiarities and adaptations for life in this dry habitat.

These experiences working in the Dry Chaco during my doctoral research influenced me as a biologist and as a person. During those years, I shared trips to the field with fellow biologists. Most of our activity took place at night. Since we studied adult amphibians when they were active and calling, we began our work at dusk and arrived at our study sites late at night. Then, very late at night, we would wade up to our waists in temporary pools full of water. We could hardly hear our voices because of the loud calling. It was like a stunning choir of frogs and toads crying out for help to find mates. These were special moments,

Gaby with two people who have inspired her work and life throughout all these years: her friend Lee Fitzgerald (left) and her husband Felix Cruz (right). Photo by Gaby Perotti.

motivators and triggers in our desire to understand more about the behavior of these amphibians.

On one of these trips, we had an experience that terrified us but also taught us two things. We were working in a pond on the side of the road. It was packed with amphibians noisily calling after a heavy rain—one of those rains that, in the Dry Chaco, appear suddenly and generate a kind of hypnotic state in these amphibians. In the midst of the darkness, lights appeared in the distance. The lights soon became a great light concentrated on us. It was a truck with several reflectors. Five people exited the truck. They were uniformed police, who pointed their weapons at us and questioned us about what we were doing, late at night. We explained that we were studying amphibians, and this calmed the policemen. Perhaps because our story was unexpected, they showed interest in our work by watching and asking questions and then finally left.

We were a little stunned by the experience, a mixture of fear and paralysis, but the resumption of amphibian calls brought us back to the work we had come to do. A couple of hours passed, and we again saw a light that was small in the distance but as it got closer it seemed like a fire that would soon swallow us. Our adrenaline levels increased. I think all four of us had the same thoughts—should we defend ourselves or run away? Suddenly, the policemen stopped the truck and called to us. We approached and saw that in a box in the back of the truck there were several dozen very large toads jumping and wriggling. The species is very common in the Chaco Seco, both in the countryside and in the towns, particularly in places that are illuminated; at night the toads are seen surrounding the lights and feeding on insects. The locals call them "Rococo Toads," respected because, according to the countrymen, they "clean up vermin." (At that time this species was designated as *Bufo paracnemis*; it is now *Rhinella schneideri*.) These policemen had thought of us and wanted to "collaborate in our study" . . . "relieve our work," so we kindly accepted and loaded the dozens of large Rococo Toads into our vehicle. We took them with us to the modest hotel where we were staying that night. This small group of Rococo Toads became part of the community of amphibians that surrounded the hotel, feeding at night on the insects that surrounded the lights and cleaning the area of "vermin."

This experience taught us two things. On the one hand, for a biologist, especially a woman, it is often complicated to work in remote places, which might not be sufficiently far from human influence, because there are potential dangers. It is important to consider this aspect and organize the work with well-established associates and manage logistics appropriately, including permissions and contacts. On the other hand, the fieldwork also taught us about the kindness of local people, their empathy, and their willingness to share all those riches that the land offers them and that sometimes city life does not allow us to appreciate or enjoy.

During those years we also grew as a family, with two boys (Luis and Lucas) who accompanied us to the field. I remember sampling tadpoles in muddy lagoons and the boys wanting to help— their boots stuck in the mud, crying and crying (they were two and four years old), carrying their small sampling nets and helping mommy (me)—while mommy cried as the tadpoles escaped in shoals to opposite ends of the lagoon and daddy tried to calm the children so we could continue the sampling. These were the first years of family life, trying to combine my passions of field biology and experiments with my family.

My work experiences are intermingled with the challenge of motherhood. The latter included leaving my small children and traveling far from home for fieldwork, sometimes healing my children's stomach pains over the phone—the typical things that working mothers must go through. I think these experiences as a biologist, combined with my role as life partner and mother, have helped me learn to ask more questions and think about them, sometimes with resolution and other times without resolution—but always thinking of possible alternatives. Combining my experiences as a woman, mother, and partner with my professional motivations, my curiosity is channeled toward new projects to discover new things. And that inspires me.

About the Author

María Gabriela Perotti is a researcher at Instituto de Investigaciones en Biodiversidad y Medioambiente (Consejo Nacional de Investigaciones Científicas y Técnicas–Universidad Nacional del Comahue), San Carlos de Bariloche, Argentina. She started her experience as a herpetologist by studying reproductive modes of anuran communities in the Gran Chaco region of Argentina. For the past 20 years her research has focused on the evolutionary ecology and conservation biology of South American amphibians. She is interested in how temperate anurans respond to a combination of naturally occurring abiotic (e.g., temperature, UVR) and biotic (predation, competition, infectious diseases) factors by studying anuran physiology, morphology, and behavior at different developmental stages. She also collaborates with Dr. Felix Cruz in evolutionary studies on *Liolaemus* lizards. Her work in collaboration with colleagues and students combines field studies and laboratory experiments.

47

ADVENTURES IN WONDERLAND

Laurie Vitt

As a child, I dreamed of spending my life in the Amazon, chasing anacondas, giant armadillos, jaguars, bushmasters, and the like. Reptile books by Raymond Lee Ditmars and adventure stories by Frank Howard Buck kept me awake at night as I imagined myself exploring tropical rainforests. As I grew up, I spent every spare minute trying to find all the reptile and amphibian species around where I lived. In doing so, I developed search images for animals that would serve me well in a career that took me to tropical rainforests of Central and South America. Those of us who have had the opportunity to live with the animals we study know how lucky we have been.

During my academic career, I chased lizards across desert sands of the American Southwest, up and down trees in the old South, in Roman quarries of Menorca, across outcrops in the semiarid northeast of Brazil, and the trees and leaf litter of Central and South American tropical forests. I collaborated with many people and published research results in international scientific journals. But that is not what I write about here. Rather, I want to take you on one of my expeditions as more or less a voyeur—a set of eyes hidden in my shirt pocket— during which I not only chased amphibians and reptiles but also found myself immersed in a wonderland of unfamiliar culture.

We begin in Boa Vista, the capital of the Brazilian state Roraima. Roraima is bisected by the north–south Rio Branco. In the south, large tracts of lowland tropical forest blend with forests of Amazonas, whereas in the north, large expanses of savanna extend into Venezuela. This and a few other savannas are known as Amazon Savannas—open areas within the bounds of the Amazon rainforest. Just south of Boa Vista, one can stand in savanna with no trees in sight, except for what appears to be a low wall to the west. The "wall" is tropical forest, and the transition from savanna to forest is abrupt. Very few lizard species occur in both, making the area of significant interest to ecologists.

When my wife (Janalee Caldwell) and I arrived in Boa Vista in 1991, Celso Morato, a biologist working for the Instituto Nacional de Pesquisas da Amazô-nia, met us and had made arrangements for us to stay in an old Catholic monastery located near the Rio Branco. We had a room with a southwest-facing outer wall, no furniture, and several sets of hammock hooks. After 15 days of exploring and collecting around Boa Vista, an opportunity to travel deep into a Yanomami reserve arose. At the time, one, especially a foreigner, could not simply enter the Indigenous reserves. Apparently, some of the Yanomamis in the Catrimani Reserve held the belief that illegal gold miners had decimated the fauna, making it difficult for the Yanomamis to hunt for food. The local director of FUNAI (National Indian Foundation of Brazil) had contacted Celso about doing a survey to see if reptiles and amphibians had declined in the area.

Of course, nothing is simple in Brazil. In addition to having FUNAI clearance, it was necessary to have the blessing of the local bishop. Our meeting with the bishop at his home in Boa Vista was surreal. First, the house was a mansion by local standards, and it was filled with beautiful furniture, rugs, and other treasures. It was like stepping into another dimension. Second, in his natural environment, the bishop seemed bigger than life, and in some respects, he was. After all, he controlled all missions in Roraima. The meeting went well, he approved of us, and we began making arrangements for the trip to the reserve. (As an aside, several years later we ran into the bishop in the Manaus airport and he appeared to be a very small man!)

We drove from Boa Vista to an east–west road north of Caracarai in southern Roraima. Sharing the Toyota with us was a nun with a Yanomami baby who had been in the hospital in Boa Vista. The road ended at a raging river. The bridge had been removed to protect the Yanomami reserve, which was on the other side. We loaded our equipment and supplies into a small boat and crossed the river, hanging on to a cable strung across it. On the other side, we were met in a pickup by Padre Guglielmo Damioli, a Macushi Indian, and a Yanomami chief. We would be traveling 100 km to Missão Catrimani. Although the road had originally been built as part of the Trans-Amazon Highway, it was a dirt road. Celso and I rode in the back of the pickup, on top of bags of rice, beans, and other supplies, with the two Indians who talked nonstop in Yanomami. As we traveled, several jaguars and mountain lions came up on the road and ran along behind the truck. At one point, a bridge that we had to cross was shaking so violently that it appeared unstable. A large log was caught on the bridge supports, and the raging high water pushing it caused it to vibrate. The Padre and the Macushi climbed down the lattice of the bridge supports to the log, carrying a chain saw, and sawed it so that the pieces would wash away. To us, it appeared extremely dangerous, but for the Padre, it was apparently routine.

About 70 km into the trip, we stopped at a small Yanomami village. Padre Guillermo had what appeared to us as an intense negotiation with one of the Yanomami men. Most of the Yanomamis in this village appeared sick and unhealthy. We learned later that the exchange was over malarial and other drugs, which were apparently in very short supply. One of the Yanomamis showed us an arrow that was about 6 ft long and weighed almost nothing. We asked where he got the arrow, and he led us down to a patch of reeds that were long, straight, and when dried, stiff. His bow was also long, straight, and extremely stiff, made from a black, dense tropical wood. When pulled back, the bow was extremely strong. We would learn later that they used several different tips on their arrows. One for monkeys was thin and straight, sharpened on both edges, and covered with some sort of toxin (probably a plant-derived toxin such as curare); one for fish had jagged edges; and one for warfare was large, flat, and sharpened on both edges. Arrow tips were mounted in split ends of arrows with a string wrapping holding them in place. Tips could be pulled straight out of the arrow's tip, allowing for recovery of arrows. For example, a monkey hit with an arrow would knock the arrow away, but the tip would remain in it. The arrow would drop to the ground and be recovered by the Yanomami.

We continued on to the Missão Catrimani, a Catholic oasis in an ocean of forest. When we arrived at the mission, we were given a small room with a screen front. The next morning, we looked out from our hammocks to see our screened window filled with faces of Yanomamis peering in at us. To us, it was like going back in time 10,000 years, and I'm sure that to the Yanomamis, we were aliens from an unimagined future. The Yanomamis had golden-brown skin and jet-black straight hair. For decoration, both adult women and female children had three sticks through their lower lips, which they could move with their tongues. They were adorned with beaded necklaces and wristbands. Beaded bands worn on their upper arms were accented with leaves or flowers. Some women wore earrings constructed of leaves. Women wore only a scant waistband and carried babies in what appeared to be a caiman-derived leather strap across their shoulder. Some women had red pigment spread in two wide lines across either side of the face. The pigment is derived from crushed seeds of the urucum (*Bixa orellana*) plant. It was also smeared on the upper arms of some men and women. Urucum and other medicinal plants are used to treat many ailments, including fevers, muscular pains, and stomach aches.

Yanomami children showing some of the typical ornamentation worn by children. Beads in the necklaces are likely imported and traded to the Yanomamis by people at Missão Catrimani. Photo by Laurie Vitt.

We stayed two weeks, spending most of our time searching for reptiles and amphibians. Much of the time, we were followed around by Yanomamis. We had set up tables outside to process and preserve specimens and to photograph each species. An old and wrinkled Yanomami, naked except for the string holding his penis erect, would stand in front of the table where my wife was working, watching for hours. We never did figure out whether he was really interested in what we were doing or was just an exhibitionist.

One day during our second week, several teenaged kids came running and shouting about something. The Padre translated to us that there was a snake in a tree. When we got to the tree, we could not see a snake. However, high in the tree we could see a bird flailing around. As I climbed the tree, I still could see no snake. When I approached the bird, I realized that it was being held in the mouth of a 5-ft Green Vine Snake (*Oxybelis fulgidus*) that was nearly invisible against the leaves. I carefully grabbed the snake (they are somewhat venomous), and it dropped the bird. I wrestled the snake into a cloth bag while hanging on to the tree. I had not paid much attention to the bird, but when I got to the ground, I realized that the Yanomamis were not interested in the snake—they wanted the bird to eat!

Several days before leaving the mission, I had remembered reading about how some Indigenous peoples were tuned in to the animals and plants around them. We set up a makeshift exam to see what the Yanomamis knew about reptiles and amphibians. They claimed that their chief was the best naturalist, so he became our "guinea pig." Because we had many preserved specimens, we set up a lab exam. At each station, we had a reptile or amphibian, and we asked a brief set of questions, such as "what is it, is it dangerous, where does it live, what does it eat," and so on. We also placed several "dupes" to see if his answers were consistent. We relied on translation by the Padre from Yanomami to Portuguese and then Portuguese to English by Celso. We discovered that the chief believed that all snakes, lizards, and frogs were venomous; banded snakes were coral snakes; tadpoles had no relationship to frogs or toads; blindsnakes (smaller species) were the same as roundworms, and they entered your body if you sat on the forest floor; upon death, your soul would enter the body of an animal, and if you were a bad singer in life, you would become a toad; and finally, that the chief knew very little about reptiles and amphibians. Whether this reflects the group knowledge is anyone's guess.

The day before leaving, Celso and I turned over several stacks of roofing that were lying around one of the buildings. The first thing I saw was a 3-ft Golden Tegu lizard (*Tupinambis teguixin*), which I put my foot on as it tried to escape. At the same time, Celso and I were trying to grab several small microteiid lizards. In the end, we collected an Ocellated Tegu (*Cercosaura ocellata*), six Underwood's Spectacled Tegus (*Gymnophthalmus underwoodi*), and the Golden Tegu underneath the last piece of roofing. As Celso and I were grappling for lizards, I felt a breeze as my wife Jan jumped out of the way. Wondering what had caught her attention, I looked down to see a huge tarantula next to where she had been standing. Jan was so transfixed on the spider that she never saw the Golden Tegu, even though it had run right under her! Little Miss Muffet. . . .

By the time we left, we felt as though we had gained many new friends and maybe even connected with our own distant past as human beings. Children played games, chased each other around, and played jokes on each other, while adult men hunted and women tended simple gardens. The Yanomami were curious, observing with great interest as we collected and processed animals. They even collected some animals for us. What initially seemed otherworldly to us, impressed us more and more as time went on. The Yanomami, isolated for thousands of years, were surviving in a complex and often unforgiving tropical lowland forest in much the same way that our ancestors did. Their most important priorities (child care, hunting, agriculture, safety) appeared no different from ours. As they did in North America 400 years ago, political and economic pressures are forcing the Yanomamis to enter a complex modern world while fighting to keep the lands to which they and their ancestors had adapted.

About the Author

Laurie Vitt received a BS and MS degree at Western Washington University and a PhD at Arizona State University. He was professor of biology at University of California–Los Angeles for eight years and professor of biology and curator of reptiles at the Sam Noble Museum of Natural History for 21 years. His research on ecology of lizards resulted in numerous scientific publications. He is coauthor of the textbook *Herpetology: An Introductory Biology of Amphibians and Reptiles*. In 2003 he received a George Lynn Cross Research Professorship at the University of Oklahoma and in 2012 a Distinguished Herpetologist Award by The Herpetologists' League. He is currently professor emeritus and curator emeritus at the University of Oklahoma.

48

IN THE *RABETA* OF THE *PAJÉ*

An Ethnoherpetological Experience

Beatriz N. Cosendey

I share with you here an experience that I regard as a milestone, whether in professional, academic, life, learning, or spiritual terms, or however else you want to categorize it. An anthropo-anthological experience.

I've been in the field of biology for over a decade, but I've always flirted with the humanities. The different cultures, peoples, myths, and origins are things that catch my attention. During my doctorate, I rediscovered ethnobiology. I say that I rediscovered it because I found, in old notes, that years before I had already been interested in this line of research. I had forgotten about it. And I fell in love all over again, with the same excitement.

Ethnobiology is a transdisciplinary science, as it involves several areas of knowledge. In addition to academic sciences such as biology, ecology, anthropology, and geography, among others, ethnobiology also involves popular and traditional knowledge. And that's where the beauty of this science resides to me.

I make this short introduction to situate my identity lane and make my feeling as palpable as possible when sharing this experience. It all began when I started to read more about this topic for an examination and became aware of the possibility of doing my postdoctorate in this area. I then spent a year learning about researchers whose work interested me. So, let's get to the story.

The year was 2019, in a still prepandemic world. I was in the last year of my doctorate in ecology, in Rio de Janeiro, looking for a project with anthropological aspects to continue my education. I had already contacted some professors with no response, and I had one on the back burner, with the note "the one I liked the most." What were the chances of receiving an answer from her? But just in case,

I sent the message. To my surprise, a day later I got a positive response. In a week, the project request. And in a month, the acceptance for a doctoral internship.

So, two months after my email, I traveled more than 3000 km to have my first anthropological experience in the Brazilian Amazon. My project was developed together with a professor of anthropology from Pará, northern Brazil, and it dealt with the effects of climate change on the local species of lizards. Our research was to investigate whether, at a time when reptiles are decreasing in population size all over the world, traditional residents have noticed any difference in the density and diversity of local species.

During this endeavor, I met Quilombolas, riverine dwellers, and Indigenous people, who told me a lot about their worlds and culture. Despite their initial shyness in talking to a "researcher," as the conversation flowed they loosened up about their knowledge. In fact, I think this is an important point to highlight. Their unanimous initial statements were "I don't know anything," "the lizards are out there," "the animals are in the backyard." From these statements we can make some inferences.

First, these lines might suggest a detachment in relation to animals. In the West, lizards are far from being part of the charismatic megafauna because they are considered disgusting or threatening. Lizards are usually surrounded by legends and stories that only increase people's distance and disgust toward them. In part, this is because people tend to fear the unknown. The second point is that the participants often did not appreciate their own knowledge. Who else could know as much about the ecology and biology of an animal as someone who has lived for generations with these animals in their backyards? However, because they don't have a technical study, or a scientific view on the subject, they think they don't know, and even more so if faced with someone from the academic world. If they only knew how much I learned from them!

All these experiences with different traditional communities were very valuable. Of course, there were good and bad times. Some experiences were really bad, as when I found myself alone among strangers who didn't seem to like my presence very much, and I depended on them to do the work, provide my meals, and even get me out of there. In these situations, everything becomes more intense, especially when communication with the outside is limited. But a community member provided me the necessary company and took me to people from neighboring villages who agreed to participate in the interviews. I came away with some results from this study thanks to him. Luckily, I always got a helping hand wherever I stayed.

But in this tale, I will focus on the last Indigenous village I visited, which made the difference for me. It stood out in several ways, such as the life experience, receptivity, people, and results. And if there is a way to materialize an imagined desire, this experience was the one that came closest. I found myself experiencing things I had always wanted when, as a little girl, I had watched reporters and researchers on television and wondered what those professionals had done to be able to be there. At that moment, I was who I wanted to be. It was I who was in the *rabeta* (canoe) of the *pajé* (witchdoctor).

Like most good times, this one came after chaos. Leaving the community I mentioned above, with few results, not knowing where to go, and with inconclusive data, I thought I had compromised my research. Luckily, however, I was helped by the city's secretary of education who offered to take me to the pajé in a nearby village. He would be in town that afternoon to pick up the school kids' lunches.

And off we went. The pajé accepted the research—the secretary was dear to them—but was reluctant to have me there. After all, what was a big city girl going to do in the village? Did she know what it was like to be in the woods? Did she have what it takes? Nonetheless, he decided to receive me. Before leaving, the secretary and I had one last thing to convince him to do: allow me to go on the rabeta. Rabeta is the local name for a kind of motorized canoe. It is a small and open wooden boat, widely used on short journeys. I don't know if out of shame or fear, but he wanted me to take a motorcycle taxi to meet him in the village while he went along the river. He ended up giving in, and that's when our (good) interaction began.

When it was time to leave, with the lunches arranged in the bow, he told me to sit down, while he pushed off the bank. But how to push that alone, with the kilos of food he had taken from the city inside the rabeta? I decided to get out to help him. Together, we managed to unhook the rabeta. He, looking at me with a mixture of confusion and fun, asked: "Do they teach how to push a rabeta in Rio?"

During the next 70 minutes in the midst of those scenic landscapes, I felt so emotional that my eyes must have sparkled. Sitting on the pajé's rabeta, I traveled a long river through the meanders of the Amazon to have my first night's stay at the Kunuwaru village. This time, I was the character in the scenes I imagined, surrounded by the woods and stilt houses of riverine and Indigenous peoples. The silence was broken only by the motor of the rabeta.

When we could see the village, I asked if there weren't Sucuri snakes (anacondas—*Eunectes murinus*) in the river. The answer was somewhat unexpected, but decisive: "Yes. Sucuri and piranha. But they have never caught anyone. You can enter," he answered calmly. That argument was enough. But how could it not be, if it was our only bathing place? Shortly after we disembarked, children ran to the bank, curious about the stranger brought from the city. After I helped with the unload, the pajé introduced me to the community. Actually, most of it, because a curious little girl had already asked me who I was the minute I landed.

We then agreed that, after I had made my lunch, we would have a group interview at Oca Cultural. Meanwhile, some children attended their class at the local school, which also taught Tupi, the language of their ancestors. The younger children, who had already had class in the morning, decided to introduce me to the area. And there I was, holding the hands of three little Indigenous girls on the right and two on the left, looking for lizards. They showed me the houses where they lived, the fruits, and what each plant was used for. I was amazed at how much knowledge those girls had, among whom the oldest must have been seven years old.

With this whole troupe it was hard not to make noise. I didn't have success in finding lizards, but I learned a lot about the local ecosystem! Not to mention the countless photos we took next to the "vinegar tree" (a plant used as a seasoning, which is "only for the strong," as one of my assistants said). It was chosen as the scenic backdrop for the girls, who asked me one by one to take pictures as they posed next to the plant.

Then it was time for the interview. I returned with my troupe of assistants in both hands. The pajé, his wife, his son (community counselor), three residents, and some children were at the Oca Cultural. We had the group interview, during which everyone talked about the local lizard species. And then they mentioned the Tamaquaré (*Uranoscodon superciliosus*). Anhhh, the Tamaquaré! Everywhere I went, it was the most mentioned lizard! There were some stars throughout these community reports, such as the white and black *osgas* (geckos), but nothing compared with the Tamaquaré.

Mentioned by all communities, and perhaps by all interviewees, it is so famous that it has a song in its honor, sung by Dona Onete (a Carimbó singer). "The tamaquaré tea/it is a very crazy tea/it is a caboclo spell/that is only available in Pará." The Tamaquaré is considered a treacherous lizard, and its main use, as a *mandinga* (witchcraft), is a tea to make the partner goofy. It is interesting to note that this esoteric usefulness comes from observing the lizard's behavior in nature. The lizard camouflages itself on tree branches near rivers in a way that its prey is not aware of its presence. So, the person who ingests the Tamaquaré tea would be just like the lizard's prey: silly.

The difference was that, this time, in addition to hearing stories about the lizard, I would finally be able to meet it! After the interview, the pajé went hunting by the river and only returned when he had a Tamaquaré in hand! I took so many photos that he said that it looked like I was the one who had cast the spell this time because, despite being skittish, the Tamaquaré stayed really still.

After the interview, I got to know the school and the crafts the community members make there. So many lovely necklaces, bracelets, and earrings made with seeds and beads, and they admired my macramé necklace! Go figure. Luckily,

I managed to take an imposing photo of the pajé holding a huge bow and arrow with which they sometimes hunted. The afternoon ended with a tag play in the river with the children—the same river with piranha and anaconda. Just in case, I found it wise not to stray too far from the children.

At night, I tried to teach the kids some macramé knots, but unfortunately I couldn't with the sewing thread we had available. Everyone had dinner together at the pajé's house, and we watched the football match on the living room television. It was also there where we slept, five or six people, including the schoolteacher, in hammocks spread from the living room beams. The next morning, I would make my way back to Belém. I still had time to say goodbye to the river and, to my surprise, also to the residents. Before I got my ride, the pajé came to thank me for my behavior during those days.

I left there with a feeling of more than accomplished mission. Besides completing the interviews, I was also congratulated by the pajé, the same who feared my stay at the beginning. And that's how my experience ended, on the back of a motorcycle, crossing communities to catch my bus to Belém. There could have been nothing more satisfying. Life experience, engraved in the soul, foundation of my stimulus to continue in this endeavor.

About the Author

Beatriz Nunes Cosendey is a Brazilian biologist who graduated in 2010 in marine biology and in 2013 completed her teaching degree in biology, both from the Fluminense Federal University. She joined the postgraduate course in ecology, finishing her master's degree in 2015 and her doctorate in 2020, at the University of the State of Rio de Janeiro. During her doctorate, she worked at the Center for Higher Amazon Studies at the Federal University of Pará with a professor of anthropology. Now she is developing postdoctoral research at the same center, to learn about the relationship between the riverside dwellers of the Amazon floodplains and the sucuri snakes (*Eunectes* sp.). She is a member of the Tropical Vertebrate Ecology Research Group at the University of the State of Rio de Janeiro; her research emphasis is herpetology. Her main interests are ecology and conservation, population ecology, ethnobiological studies (mainly ethnoecology and ethnoherpetology), and science communication.

PARTING THOUGHTS

For centuries, scientists and naturalists have documented the world's biodiversity and studied the basic biology of organisms in the field. Increasingly, in addition to these two lines of research, scientists study ecological interactions of plants and animals in their natural environments, often incorporating field experiments. Species new to science are still discovered every day. Fieldwork is crucial. We use field observations to form scientific questions, make predictions, and test hypotheses as we strive to understand the natural world. Unfortunately, however, less emphasis and fewer resources are currently directed to fieldwork than in past decades. Instead, studies aimed at the molecular and cellular levels often receive the lion's share of funding, while field studies are left with the dregs or left unfunded altogether. Field studies often require greater time investment and yield fewer publications as compared with other types of biological research, for example, some kinds of physiological studies conducted in a laboratory under controlled conditions. Number of publications and success in bringing in grant funds influence an academic's ability to achieve tenure and promotion. The result is that field biologists sometimes find themselves at a disadvantage as compared with their colleagues in this regard.

But the fact is that fieldwork is as critical as ever, and perhaps even more so. With up to a million species threatened with extinction (United Nations 2019), we are running out of time needed to document and to understand the basic biology and ecological interactions of much of the world's biodiversity. Human-caused habitat modification and destruction worsen the problem. The worlds pictured in many of the essays in this book have changed since the time the fieldwork was undertaken. In my case, Santa Cecilia, once a small Quechua village scattered along the Río Aguarico, is now a community of over 7500 residents and little forest remains nearby. Worldwide, natural areas are being converted to cattle pastures, agricultural croplands, and residential areas, all to satisfy the needs of one species—*Homo sapiens.* We are losing wild places and the animals that call those places home.

Through field studies, we can better understand the basic needs of animals and be in a stronger position to fight for protection of species and preservation of their habitats. We will always have a need for fieldwork. And we need more field biologists.

Reference

United Nations. 2019. UN report: Nature's dangerous decline "unprecedented"; species extinction rates "accelerating." https://www.un.org/sustainabledevelopment/blog/2019/05/nature-decline-unprecedented-report/.

Acknowledgments

Thanks to Kitty Liu for believing in this book from the moment I proposed it to her. I appreciate the support from Jackie Teoh, Susan Specter, Eva Silverfine, and the rest of the team at Cornell University Press.

Many thanks to Maureen Donnelly, William Lamar, Alan Savitzky, and especially Karen McKree for reading parts of my prose and for offering helpful feedback. Thanks to Alison Davis Rabosky for her superb rendering of the map indicating our field sites highlighted in this book.

Finally, my deepest thanks to all the contributors for their enthusiasm for the project and for their compelling storytelling.

Index

Page numbers in **bold** refer to figures.